T0226095

Cities Research Series

Series Editor

Paul Burton, Gold Coast campus, Cities Research Institute, Griffith University, Southport, QLD, Australia

This book series brings together researchers, planning professionals and policy makers in the area of cities and urban development and publishes recent advances in the field. It addresses contemporary urban issues to understand and meet urban challenges and make (future) cities more sustainable and better places to live. The series covers, but is not limited to the following topics:

- Transport policy and behaviour
- Architecture, architectural science and construction engineering
- Urban planning, urban design and housing
- Infrastructure planning and management
- Complex systems and cities
- Urban and regional governance
- Smart and digital technologies

More information about this series at http://www.springer.com/series/16474

Scott Baum

Editor

Methods in Urban Analysis

 Springer

Editor
Scott Baum
Griffith University
Brisbane, QLD, Australia

ISSN 2662-4842　　　　　　　ISSN 2662-4850　(electronic)
Cities Research Series
ISBN 978-981-16-1679-2　　　ISBN 978-981-16-1677-8　(eBook)
https://doi.org/10.1007/978-981-16-1677-8

This Springer imprint is published by the registered company Springer Nature Singapore Pte Ltd.
The registered company address is: 152 Beach Road, #21-01/04 Gateway East, Singapore 189721, Singapore

Foreword

I am delighted to introduce the second book in our Cities Research Series. As more and more of us across the globe live in urban areas, whether they are called cities, towns, conurbations or urban villages, it is increasingly important that we better understand the different experiences of urban life and the forces shaping these places. Urban analysis is a useful term that captures the range of techniques and approaches to studying these experiences and forces. In the past, we might have become preoccupied with arguing over the meaning and specificity of the term 'urban', but we now recognised that towns and cities take many different forms and that these change over time, sometimes disconcertingly quickly. We also now recognise that places, whatever they might be known as, are connected both physically and, increasingly, digitally. The recent pandemic has shown vividly some of the consequences of this physical connectivity as viruses like many of us now travel around the world at a pace previously unimaginable. And greater digital connectivity has undermined the tyranny of distance for many aspects of our lives so that we can work, learn and socialise together across continents if need be, even if we still value physical connections and suffer when we are denied them.

As the chapters in the collection demonstrate, many of the techniques and approaches to researching urban experiences and phenomena are not exclusive to the urban realm. The principles of systematically reviewing a body of published work apply whatever the topic or focus and good survey research about urban issues are not significantly different from good survey research about issues found only in rural and remote areas, although there may be a different set of practical and organisational challenges to conducting surveys well in different settings.

While a capacity for rigorous urban analysis is the foundation for good planning practice, many other professionals, policymakers, practitioners and activists will be better placed with these skills and competencies. This collection of practical and thoughtful accounts of various forms and techniques of urban analysis offers an up-to-date resource that we believe will be useful to all of these groups. Critical and reflective practice as urban analysts will strengthen our capacity to better understand urban places and to help them become better, more resilient and fairer places.

If you would like to propose a contribution to the series, please contact me directly.

March 2021

Prof. Paul Burton
Series Editor
Cities Research Institute
Griffith University
Southport, QLD, Australia
p.burton@griffith.edu.au

Contents

Why Urban Analysis?

Scott Baum

Abstract This chapter introduces some of the basic principles and practices of undertaking research in urban analysis or urban planning courses. It begins by asking the question: 'What is urban analysis' and discusses the reasons why we might undertake urban analysis or planning research. It then introduces the reader to several basic research concepts that form the basic knowledge set covering most forms of research undertaken in the field. This includes introducing the reader to the different purposes of research, different methodological approaches and data sources, as well as to the important question of ethics in research.

1 What is Urban Analysis?

Chances are, that if you are reading this book then you are interested in cities and urban settlements, or you are doing a course in planning, geography or a related field. As such, you may already have an idea of what urban analysis is. In this book, we treat urban analysis as being the empirical study of cities and urban life. Moreover, we consider urban analysis as being a collection of approaches, methods and procedures that aid in our understanding of cities and urban life. Understanding the methods associated in doing urban analysis is certainly useful if you are an academic or are thinking of a job as an academic or professional researcher. But understanding the methods of urban analysis is also important if you are going to be working in government as a planner, in a non-government organisation as an advisor or any number of other jobs. In fact, given that a lot of social life we witness happens in cities or urban settings, it is likely that even if you never work in a job with requirements for either conducting or understanding research and its outcomes, knowing the how-to and why of urban analysis will help you better understand the world we live in.

In a more formal way urban analysis can be considered from two interrelated positions: an academically oriented position that focuses on theory building or testing

S. Baum (✉)
Griffith University, Brisbane, QLD, Australia
e-mail: s.baum@griffith.edu.au

© Springer Nature Singapore Pte Ltd. 2021
S. Baum (ed.), *Methods in Urban Analysis*, Cities Research Series,
https://doi.org/10.1007/978-981-16-1677-8_1

and is generally referred to as basic research and a policy-focused position that is applied in nature and generally focuses on solving a particular problem. The basic and applied ends of the research spectrum are not mutually exclusive as the foci and interests of both often converge, as implied above, on wanting to better understand the urban world we live in. Indeed, the methodological toolbox used by both basic researchers and applied researchers is the same although the actual approaches used in any single project may differ.

Many of the examples given in this book will be focused on applied research. Applied urban analysis focuses on the how, what and where of urban areas with the goal of developing a nuanced understanding of a given issue for use in the policy making area. If you are undertaking applied urban analysis, then you are going to focus on the identification and answers to problems as they occur within cities and urban areas. You might, for example, investigate the most preferred location for a stadium, seek to answer questions about public transport usage or you might be engaged to provide answers to questions around community satisfaction in the wake of a range of social issues. All of these might be of interest to an applied urban researcher. And they would all follow a formal and methodical approach to arrive at a well-considered answer.

This leads to another point regarding urban analysis. While it may not necessarily always be the case in practice, urban analysis should follow a well-thought-out methodological path or framework. Regardless if you are undertaking the basic research or applied research, good urban analysis needs to abide by a carefully considered and systematic recipe. Issues will include

- Defining the scope of the research
- Identifying variables
- Considering different levels of analysis and causal relations among variables
- Identifying potential sources of data
- Estimating data collection needs
- Taking an inventory of the resources needed to proceed
- Considering the ideal (most rigorous) research design
- Identifying potential findings
- Targeting your audience and determining how you will present your findings

Each of these issues has a long-established body of appropriate protocols and approaches, many of which will be elaborated on in this book.

2 Why Do Urban Analysis?

Let's delve a little more into the why of urban analysis. As I have noted above, even at the most basic level the simple reason we want to know about urban analysis is to have a better understanding of what is going on around us as we go about our daily life in the city. Still not convinced? Think about this. Often, after we've lived in a city for extended time, we think we have a good idea of the structure and processes that are underway and shape the city's social, economic, demographic and environmental fabric. For example, residents of a city can usually point out the socio-economic divides that characterise a particular place, what neighbourhoods are most desirable to live in, and why and what is the major physical characteristics of their local area.

As an experiment, I often get my students to list the things they know about their city or neighbourhoods' broad social and economic fabric. I ask them to consider how their city is compared to the wider state or regional average in terms of incomes, family types, levels of crime etc. I ask them if they think they know their city well then get them to draw a diagram or infographic to illustrate their understanding.

I then present them with an infographic illustrating some of the key factors that determine the social and economic shape of the city. Once presented with this, the student often argues that my analysis benefits from the use of official census statistics or some other data that they didn't have access to.

I point out that the difference between the two styles of analysis is that one is relying on their personal and individual experience of the city or neighbourhood they live in, while the other relies on data collected by methodical and systematic research approaches. My goal is to illustrate the importance of urban analysis (and all research for that matter) and to use established methodologies and approaches to develop sound evidence which enhances our understanding of cities and urban areas.

Let's take another example. If you were to select a suburb at random, could you provide information about the level of satisfaction that a person residing in that suburb might have and could you provide some reasons to account for that level of satisfaction? Issues around residential satisfaction are one area of urban social life that often comes up in the academic literature [1–3]. If you read some of this literature you will encounter some theories or hypotheses about urban life and residential satisfaction that might provide you with some broad ideas, but you still couldn't really answer the question. Certainly, you could have a look at the physical characteristics of the suburb and that might give you some idea. Beyond that you might simply have to rely on your own views or knowledge. If you did this, then you would probably be missing out on a whole lot of issues and would have only a very superficial (and possibly widely incorrect) view of the subject matter. Again, this is an illustration of the need to have a good grasp of the methods and approaches that are the backbone of urban analysis.

2.1 Systematic Versus Non-systematic Knowledge

The point I am alluding to above is that the way we understand things is usually made up of knowledge gained from several different sources. This applies to both our understanding of urban areas as much as our understanding of general everyday life. In understanding the social shape of the city for example, we might base our view on our personal experience of where we live and the surrounding suburbs or on what our friends told us about where they live. Within the realm of research our knowledge about things gained from personal experience is placed within the orbit of non-systematic sources of knowledge. Babbie [4] refers to this as ordinary, non-scientific inquiry. These non-systematic sources of knowledge are 'those things we know as part of the culture we share with those around us' [4, p. 5] and include knowledge gained through common sense (everyone knows that it is so), intuition (I just know that it is so), beliefs (based on personal conviction), tradition (it has always been that way), personal experience or authority (the word of an 'expert'). Obviously, some sources of non-systematic knowledge are more credible than others. However, as I pointed out above, only relying on these types of knowledge to inform reality can result in a piece-meal view of the society in which we are investigating.

Systematic sources of knowledge involve methods and approaches for conducting empirical research that comply with rules that specify objectivity, logic, and communication and the link between research and theory. With systematic sources researchers follow a set of criteria to make informed judgements about a particular issue.

So why systematic sources of knowledge? Within the realm of urban analysis, we often want answers to important questions that impact on a whole city or a particular community. This often involves having considered input into policy. Within the world of policy, systematic sources of knowledge contribute towards the evidence base. For instance, in the residential satisfaction example used above, by following a systematic approach to answering the question (i.e. undertaking a sophisticated social survey with a large robust sample) we could provide much more considered input into policy than if we were simply to rely on personal experience.

3 Research Basics

The main goal of this book is to provide a guide to undertaking aspects of urban analysis. Its aim is not to provide answers to all methodological issues, but rather to highlight some of the most commonly encountered. However, prior to diving into the main material, it is useful to consider a number of fundamental research issues that establish the context for the material presented in subsequent chapters. Here I want to discuss the purpose of research, methodological approaches, sources of data and ethical issues.

3.1 The Purpose of the Research

We have already discussed the idea that urban analysis can be considered as being basic research (theory testing or building) or applied research. Beyond this distinction, a researcher must also decide on the purpose of the research they are undertaking. Understanding the purpose of the research is often an important founding step in developing a research idea (see chapter "Research Questions and Research Design").

At its most basic level, research can be viewed as a simple process of exploration. Researchers may undertake a program of exploratory analysis as part of an early pilot study or as a part of the background to a larger study. Researchers will use exploratory studies when they are new to a particular subject area and wish to survey the existing research field. As an example, let's suppose that you decide to undertake a research project on gentrification as part of your university degree. In doing so, you may want to know a bit more about the subject area before you begin. You could carry out a literature review (in essence an exploration of the research and policy literature) and you could perhaps ask or interview some 'experts' about some of the issues. You might be interested in exploring what is gentrification, what studies have already been done, is it an issue for concern in your city, and if so, what are some of the policy questions or implications. By exploring the issue, you are likely to come up with many unanswered questions and this exploration may provide important information for moving your research forward.

A significant body of research in urban analysis is given over to providing descriptive analysis of an issue or problem. Descriptive research uses methods to systematically describe situations, events or behaviours. It asks the what, where, when and how questions. For instance, a descriptive analysis of an urban area's demographic makeup might involve providing basic tables, graphs or maps that show how the age structure differs between suburbs or how certain zones in a city have higher levels of renters than homeowners. Descriptive research studies often make use of official statistics such as population census as the work of demographer Bernard Salt [5] on population change or my own work on community typologies [6] illustrates. Research that involves policy evaluation, say of urban planning outcomes, is also likely to be descriptive, as is research tagged as social, economic or environmental impact assessment that aims to determine the likely impact of planned changes.

More complicated than the descriptive research is the research that aims to gain understanding of a particular issue and explain outcomes or processes. Explanatory research aims to answer the why questions. For instance, while a descriptive analysis of residential mobility might show who moved and where they moved to within a city, an exploratory analysis would also look at the reasons why particular groups of households move and the reasons and processes behind their mobility choices [7, 8]. Here a researcher would enlist theories or conceptual frameworks to guide their research and search for statistically significant correlations or associations between variables.

3.2 Methodological Approaches

Within the urban analysis literature as well as more broadly within the social research literature, there are two broad methodological approaches used—quantitative and qualitative. Quantitative approaches utilise numerical representations of observations in order to answer a research question. For example, the residential mobility example described above might use data from a survey to show numerically how many people move and the stated reasons for moving. This data may be displayed in a tabular form, as a graph or might form part of a complex mathematical equation.

In contrast to quantitative approaches, qualitative approaches utilise non-numerical representations to answer research questions. Here information might come in the form of text, speech or some visual format such as photographs. We might, for example, find out about residential mobility by undertaking in-depth interviews with household who have recently moved and use their stories to answer the question why people move.

While it is often the case that textbooks represent these two approaches as opposite ways of approaching research, and academics often favour one approach over the other, it is the case that the two approaches are not mutually exclusive. Each have advantages and disadvantages and follow some different rules. While a quantitative approach can offer a broad overview of a topic using, for example, a complex statistical process, a qualitative approach can offer a more nuanced understanding of an issue. In this way, quantitative approaches can provide the broad skeletal view of an issue, while the qualitative approach adds some skin to the bones. Hence large research projects may enlist both approaches to provide a well-rounded analysis of the issue. In this case the research will be engaged in a mixed-methods approach. In formal terms a mixed-method approach represents research that involves collecting and analysing, both quantitative and qualitative data in a single study.

3.3   The Source of Data

As we will see, a large component of this book deals with different ways of approaching urban analysis and in particular different ways of collecting and gathering data. The data one might use in undertaking urban analysis can be classified as primary or secondary. In any single project the researcher might use secondary data, primary data or a mixture of both. Primary data are the data that has been newly collected by the researcher, most often for a specific research purpose. Secondary data is the data that has been already collected and can include survey data that another researcher has collected for another purpose or official data from government organisations. As we will see, official government statistics in the form of population census or other administrative datasets are widely used in the area of urban analysis and have become more easily accessible due to high-speed internet and high-powered computing solutions. Both approaches have their advantages and disadvantages and the sources used in any single project will depend on many factors, including cost, time and personal preference.

4 Ethics in Research

If undertaking research involves following a well-planned research path or framework in order to form systematic sources of knowledge, it is equally about following a range of rules that govern the ethical conduct of research. We often hear about the need to undertake ethical research on animals, but many of the same rules, plus some additional ones, pertain to the conduct of social research. In essence, ethics brings us into the realm of values in the research process. Here we ask questions such as how should we treat the subjects of our research? or Are there types of research or types of research questions in which we should or should not engage in? Large chapters of social science research textbooks are filled with discussions around these issues (see, for example, Babbie [4] for an extensive discussion) and all research institutions, universities and many funding bodies have detailed research ethics handbooks and guides that set out in detail the main ethics principles that researchers are expected to abide by.

Ethical principles are simply the set of values and standards used to determine appropriate and acceptable conduct at all stages of the research process. They are a set of moral and social standards that includes both prohibitions against and prescriptions for specific kinds of behaviour in the research setting. Depending on the types of research being undertaken, an individual or a team of researchers are faced with four broad ethical principles:

- Harm to participants
- Informed consent
- Invasion of privacy
- Deception.

When conducting research, one of the first issues to consider is, will the planned research approach or the findings from the research cause harm to the participants of that research. This principle has arisen due to a number of infamous research projects that, at the time, resulted in significant harm to those involved. These included the experiments by Nazi scientists on prisoners of war and the famous Milgram obedience studies [9]. Harm can be interpreted as physical harm, but often in the types of research conducted under the umbrella of urban analysis harm may be interpreted as including psychological, emotional, legal, social or financial harm. In cases where there is the potential for harm, the researcher must take steps to ensure that appropriate harm minimisation strategies are put in place. Strategies may include changing the way the research is conducted (i.e. using a different data collection method or approach) or providing participants with the opportunity to debrief following the research.

In research where the involvement of human subjects is necessary, the researcher or research team needs to obtain informed consent and ensure that the participant is not coerced to participate. The notion of informed consent requires that a participant be given full information regarding the research. This information includes details about the research itself, what participation in the research requires, any risks or harms associated with the research, what will happen to the information that is collected and information regarding how to contact the research team and/or the appropriate ethics committee or group. The participants also need to be reassured that they can withdraw from the research process at any time without penalty. While it is usual in face-to-face or paper-based surveys to obtain written consent (see chapter "Conducting Survey Research"), consent may also be oral or even implied, depending on the nature of the research approach.

Any participant in a research project must be guaranteed that their privacy will be respected. Depending on the type of research being conducted, it may be necessary for participants to disclose private or sensitive information to the researcher. The researcher should not use names or identifying information on questionnaires, and where responses can be linked to names, the researcher must ensure names and data are kept separately and only accessible to research team. One area where confidentiality and anonymity sometimes become a problem is when the researcher undertakes in-depth interviews with stakeholders and then provides reports on these interviews. In such cases, it is necessary to use pseudonyms for participants (i.e. informant A said....).

A final ethics principle relates to deception. Deception in research can take many forms from not giving a participant full information about the project or not informing a participant they are being studied to falsifying research findings. Within the research world there is generally a negative reaction to deception usually due to moral distaste (i.e. it's bad to lie) or that deception may bring the discipline or institution into disrepute. Depending on the research topic and the approach taken, it may not be possible to completely remove all forms of deception (i.e. in the case of research involving covert observation), but the researcher must ensure that all steps are taken to reduce or limit deception.

Depending on the type of research undertaken and the associated ethics issues, a researcher, prior to commencing their data collection, must apply for clearance from an appropriate body. This is usually a research ethics committee or an institutional review board (in the United States). The ethics committee (usually a university-based committee, but could also be attached to a funding body) is tasked with reviewing the proposed research, identifying any issues that are attached to the conduct of the research and providing guidance to the researcher regarding the most ethical way to conduct their research. They are also the first point of contact in the case where ethics principles have been ignored or broken.

5   Conclusions

This chapter has set the scene for the remainder of the book by presenting a number of key points to consider when undertaking an urban analysis project. The task of the remaining chapters is to assist the researcher in undertaking an urban analysis or planning research project. It is not meant to provide an in-depth treatment of the whole of the research methods field. The aims here are more modest and limited to providing a guiding framework or context for asking research questions and implementing a research plan in urban analysis or planning research. Each chapter deals with a particular aspect of research, whether that be research design, conducting a literature review or collecting and using data. To help readers expand their knowledge beyond the book, each chapter also provides a list of resources that provide further information.

Key points

- Urban analysis and planning research is a broad field that focuses on helping to understand the way cities and urban areas function, how they develop and how the networks of people, places and organisations all work together.
- Like all social science type research, research that is undertaken in the urban analysis and planning areas should be guided by a range of research basics.
- There are several possible broad approaches that may inform your research including decisions about the purpose of your research and the type of approaches you might take.
- Like all research, the research that you might undertake in the field of urban analysis or urban planning must comply with clearly set out ethics principles.

Further information

For those wanting to get further information about research basics and social science research more generally, see:

- Martyn Denscombe (2009). Ground rules for social research: Guidelines for good practice. McGraw-Hill Education (UK).
- Elisabete Silva, Patsy Healey, Neil Harris and Pieter Van den Broeck (Eds.). (2014). The Routledge handbook of planning research methods. Routledge.
- Alan Bryman, (2016) Social research methods. 2016: Oxford university press.

References

1. Baum S, Arthurson K, Rickson K (2010) Happy people in mixed-up places: the association between the degree and type of local socioeconomic mix and expressions of neighbourhood satisfaction. Urban Stud 47(3):467–485
2. Boschman S (2018) Individual differences in the neighbourhood level determinants of residential satisfaction. Housing Stud, 1–17
3. Abass ZI, Tucker R (2018) Residential satisfaction in low-density Australian suburbs: the impact of social and physical context on neighbourhood contentment. J Environ Psychol 56:36–45
4. Babbie E (2015) The practice of social research. Nelson Education
5. Salt B (2004) The big shift. Hardie Grant Publishing
6. Baum S, O'Connor K, Stimson R (2005) Fault lines exposed: advantage and disadvantage across Australia's settlement system. Monash University ePress

7. Coulter R, Scott J (2015) What motivates residential mobility? Re-examining self-reported reasons for desiring and making residential moves. Populat Space Place 21(4):354–371
8. Morris T (2017) Examining the influence of major life events as drivers of residential mobility and neighbourhood transitions. Demographic Res 36:1015–1038
9. Blass T (1991) Understanding behavior in the Milgramobedience experiment: The role of personality, situations, and their interactions. J Person Soc Psychol 60(3):398

Research Questions and Research Design

Scott Baum

Abstract This chapter introduces readers to the initial steps of designing a research project and sets out the major considerations that need to be addressed in research design. It guides the reader through issues around developing a research question and research topic, including how a researcher might come up with a good idea for a project. The chapter briefly discusses the essentials of a good literature review and provides a consideration of the important link between theory and research and the need to link your research design and research strategy.

1 Introduction

The chapters in this book present details regarding an array of different methods and approaches that might be used by researchers in the urban analysis and planning research fields. An important component in any research undertaking is developing a good research question and building a research design that will help you answer this research question. Many of the chapters touch on issues around research design specific to their particular topic. This chapter sets the scene for undertaking an urban analysis project and in particular looks at the steps, processes and pitfalls in designing a research question and subsequently planning a research project.

While researchers often view this planning phase as a boring part of the research process, not undertaking early planning is like trying to build a house without a plan. It doesn't matter if you are developing a project for a 1 million dollar research grant, a Ph.D. thesis, an undergraduate research paper or a consulting project, good initial planning is the key to a successful outcome. While it is certainly the case that plans do change during the life of a project, beginning with an outline that includes a well-designed research question or questions, a good understanding of the relevant literature and a logical methodology are essential.

One thing every new researcher soon learns is that the research process is a messy business. From start to finish, the process throws up false-starts, the potential for

S. Baum (✉)
Griffith University, Brisbane, QLD, Australia
e-mail: s.baum@griffith.edu.au

© Springer Nature Singapore Pte Ltd. 2021
S. Baum (ed.), *Methods in Urban Analysis*, Cities Research Series,
https://doi.org/10.1007/978-981-16-1677-8_2

massive side-tracks, mistakes and sometimes a complete rethinking of the research itself. But this is okay. In some ways this is part of exploring the topic you are interested in. So, although the accounts in this book suggest an ordered logical process, it is not always so. A survey might not work the way we hope, despite all the planning and piloting in the world. Analysis of your data might not work out the way you first thought. True, a piece of research may go exactly to plan, and this is what is often portrayed in the research reports and journal articles we read. But often reports and journal articles are written in a way that only focuses on how the actual findings were produced. This is not deceptive but is just that methodological discussions in reports and articles tend to follow a particular framework. At the end of the day, as long as you follow the accepted approaches for your particular area, you should be confident that the research you carry out is robust.

2 Developing a Research Topic

Very often you will be given a research topic by your lecturer or project leader, or in the case of a consultancy, your client, but you may also be called upon to develop your own research topic. The question I often get asked by students doing an undergraduate research thesis is 'how do I come up with a topic and/or a research question'?

The reality is that your research ideas can come to you in many ways and during many situations. As a student, you might decide on a broad area of research that you looked at in a class or course you have already taken, or your broad topic may come about from an interest you have outside of formal classes. For example, you might be broadly interested in environmental sustainability or social class or you may have taken a really interesting course on transport planning or demography. Or, you may have read about an issue in the local paper that sparks your interest. Any of these could be your starting point with which to explore a possible research idea. The trick is you have to start somewhere, and this is as good-a-place as any.

Many of the broad topics for classic studies in the urban research literature have come from the observations made by researchers about the cities and societies they live in. Often, they focus on a particular issue. The Chicago School Urban Sociologists, for example, developed a significant body of research (and theory) on the back of issues such as immigration and industrial change which they considered where impacting and transforming the face of Chicago during the early 1900s [1]. Authors such as Burgess et al. [2] in their book 'The City' identified theoretical and empirical issues associated with social and physical redevelopment of the inner city and established a significant research legacy. For the Chicago Urban Sociologists, the city was a large human laboratory which allowed them to explore the vast changes that were a key to understanding the way cities evolved and grew, where different groups settled and how change was reflected in the social and economic structure of cities. In a similar way the earlier research by Booth [3]—which went on to influence the research by the Chicago school—grew from a need to investigate a particular issue, the pauperism of UK citizens in London in the wake of the Industrial Revolution. One of the most important products of the Booth's work is the maps of London, which used coloured street maps to indicate the levels of poverty and wealth.

Moving to more contemporary studies, the work by Australian sociologist Arthurson [4] into the effectiveness of social mixing as a policy response to poverty grew in part from her time in the South Australian Public Service and her interest in housing and housing problems. My own work on employment vulnerability was driven by an interest in the potential impacts on urban communities following the Global Financial Crisis. In particular, I was concerned that while we knew about disadvantaged communities, we didn't have a particularly clear idea about what communities might be at economic risk in the wake of the Global Financial Crisis that began in 2008 [5].

When developing a research question, it is sometimes useful to develop a concept map of your research ideas. A concept map is a diagram that depicts suggested relationships between concepts and ideas. In developing a research topic and a research question, concept mapping helps you focus on a particular subject area and can help organise and structure what you know about a topic [6]. In a concept map, each word or phrase connects to another, and links back to the original idea, word or phrase. Concept maps are a means of developing your initial ideas about a research area. They help you identify what you know and what you don't know and help you identify the key concepts you are trying to research. Concept mapping can also help you focus down a broad research topic into a focused research question and help understand the main concepts. Developing your concept map can be done using just a pen and piece of paper or you can use one of the available software packages such as ClickUp (www.clickup.com) or mind meister (www.mindmeister.com).

So, as you can see your ideas for a research project can come from many sources, developing a strong research question or set of questions is the next step.

3 The Research Question

You may have decided on a broad research topic, but this is only the start. The defining moment of your research project will be when you nail down a research question. A research question provides a precise account about what the researcher wants to find out. It is worthwhile taking some time and making sure it is well crafted. Without a doubt, a well-crafted research question helps to ease the way for the remainder of the research project. Good question/s help to guide your review of literature, provide a guide to the types of data collection you might undertake (primary or secondary) and whether you might use quantitative approaches, qualitative approaches or a mixture of both. Very importantly, having a clear research question stops you from going off on a tangent—having your research questions front of mind help to pull you back to reality if you go off track.

Research questions can be grouped into a number of categories [7, 8] according to the main focus. Denscombe [7, pp. 11–12] suggests research questions can:

- Forecast an outcome—what will happen in the future?
- Explain the causes and consequences of something—why do things happen?
- Criticise or evaluate something—how well does something work?
- Describe something—what is it like?
- Develop good practice—how can it be improved?
- Empower—how can it help those being researched?

During the early stages of development, I often get my Ph.D. students to produce a diagram containing their research questions, linked to the types of literature and key words they will search for and the specific research approaches they will follow (this can be an extension of the concept map). This helps them to keep on track and know how their research plans are linked back to answering their questions. In this way, having good research question/s is important to the whole process. It helps to keep you focused and it is important to recognise that developing a research question is not a waste of precious time you could be spending on other parts of your research. But how do you come up with a good research question and what makes up a good question?

Australian sociologists Bouma and Ling [9, p. 14] suggest two broad properties that make for a good research question. First, a good research question must be limited in scope. That is, it is impossible for any research question to answer everything about a given topic, so while you might start off with a set of very broad questions, you will need to narrow them at some stage. This narrowing is very often the light bulb moment in a research project. Secondly, a good research question has to be researchable. Can you actually undertake research to answer the question/s. Is it feasible to undertake the research in terms of cost, ethics or your own expertise? Can you find the necessary subjects to interview? It is pointless designing a research question if it can't be feasibly researched. For example, it would be very difficult from an ethics point of view to undertake research into illegal graffiti in urban areas if your question involved documenting illegal graffiti artists at work. Similarly, let's say you wanted to investigate the changes in poverty at the suburb level annually across a 40-year timeframe. Your research question might be 'how has intra-city suburban poverty changed between 1978 and 2018'? Such a question would be almost impossible to investigate simply due to the lack of consistent data.

Attributes of a Good Research Question
1. Focused and specific.
2. Asks about the relation between the variables.
3. Is a question—not a topic.
4. Can be answered by a research process—not an argument based on values.
5. Is feasible and ethical to answer via a research process.
6. Often informed by theory.
7. Can be 'blue sky'-cutting edge, exploring new areas.

8. Can involve revisiting topics already researched (but not directly answered).

So how do you generate a research question? Researcher White [8] provides a great overview is his aptly named book 'Developing Research Questions', but I can tell you from experience that it can sometimes be hard to move from a broad topic or area of interest to a set of focused and achievable research questions. Depending on your topic, there can be so many issues that interest you, but at some stage you need to settle on a focused succinct question or set of subquestions.

I often suggest that my students begin by generally reading around the topic. What is the current literature in your study area? Does this literature lead to questions that remain unanswered? Is there a hole or gap in the research area that hasn't been explored? I also get them to list out various concepts attached to their particular broad topic as sometimes thinking about a concept (i.e. community disadvantage) may help lead to focusing down on a particular research question.

From my own research for my fourth-year thesis, I knew that I wanted to look at something from an urban sociology or human geography standpoint focusing on social and economic disadvantage. This was a broad topic area that married interests developed during my Bachelor of Economics and also through Sociology courses I took as part of my studies. I went away to the library and spent time looking at various journal articles and happened upon an article written by Sydney Economist Stillwell [10, p. 3]. The abstract for the paper read:

Sydney has been experiencing rapid structural economic change in the 1980s. This has influenced conditions in the labour market and the housing market, resulting in significant redistribution of income. One clearly observable dimension of these changes is growing spatial inequality between the component local government areas. Analysis of the 1981 and 1986 Census data illustrates these trends. It is argued that, because spatial inequity is largely a manifestation of national economic forces, its redress requires a reversal of macroeconomic policies.

Eureka!! I had managed to narrow down my focus and develop my research question. At the time, I was living in Adelaide, a city that had been significantly impacted by de-industrialisation. My broad research question became: 'who are the urban winners and losers in the period following de-industrialisation in Adelaide'? So essentially, my research question involved the replication of a previous study to an unstudied area or population. The outcome of that research can be seen here [11].

Developing a research question is also linked to the aims of our study. What is it that we want to achieve from our research? Are we simply going to describe a partic-ular phenomenon, or are we leaning more towards trying to explain a phenomenon? Being clear about the aims of your research allows you to develop a good research question and also begin understanding how you will answer your research question/s.

4 Literature Review

At some stage, early on in your project design, you will have to undertake a review of the literature. Conducting a literature review serves several purposes within the context of a research project. Early on in a project, a scan of the literature can help you avoid 'reinventing the wheel' and help you identify what is known and what is not known about a particular topic. If you are unfamiliar with a topic area, a literature review can help you identify relevant theories, research frameworks or concepts, or help you identify features of your research project that you hadn't realised were important. A literature review can also throw open your thinking about the most appropriate methods to utilise when attempting to answer your research questions. In reviewing the literature, you should try to be comprehensive, but remember to be selective at the same time. It is also important, where appropriate, to keep your review as current and up-to-date as feasible. The nuts and bolts of conducting a literature review can be reviewed in a number of publications [12] and the chapter in this book by Catherine Pickering, Malcolm Johnson and Jason Byrne provides an in-depth discussion and instructions for conducting a quantitative literature review.

Many university libraries conduct training on searching for literature and undertaking literature reviews. In doing so, they will point you towards the most appropriate sources of literature for the particular field you are interested in. In general, there is a wide array of possible sources of literature with some being more relevant than others. With access to the Internet, it is possible to efficiently search for appropriate literature through simple keyword searches (i.e., through Google Scholar). It is important, however, to think carefully about your keywords as this will impact on the types of literature you will find.

Common sources of literature can include: Peer-reviewed articles in scholarly journals: These are journal articles that have been subject to review by other academics or other experts in a field. Most often these are 'blind' peer-reviewed, meaning that the person reviewing the manuscript does not know the identity of the academic who submitted the manuscript for publishing consideration. Very often these are the most valuable sources of literature as they provide a number of important pieces of information regarding theoretical frameworks, appropriate methodology and research design.

Conference proceedings: These are akin to edited book collections in that a conference organiser will pull together papers from a conference into a published body of work. Conference proceedings can be peer-reviewed or non-peer-reviewed and are sometimes a good source of current research thinking.

Books or edited books: Another source of academic literature are books and collections of edited chapters. These are useful as they may contain a significant body of research in an area or topic and in the case of edited books may provide many different points-of-view on a topic. Depending on the publishing process these can be peer-reviewed or non-peer-reviewed.

Research reports: One literature source that emerges from outside of mainstream academic publishing are research reports and policy documents. Government and non-government bodies undertake research on a range of different topics and publish the findings as research reports. Within Australia for example, the Australian Housing and Urban Research Institute (AHURI) publish a wide range of research reports covering topics in the housing and urban analysis fields. Government departments such as the Bureau of Statistics publish research reports as do policy think-tanks.

Internet sites: The use of Internet sites as a legitimate source of literature has grown over time. While many sites provide useful information, others do not adhere to strict review or editorial policies (i.e., Wikipedia) and the information these sites contain should be avoided. They may, however, point you towards other useful information and can sometimes help in the beginning stages. One useful site for getting ideas is 'The Conversation' a web site containing short articles by academics often focusing on contemporary issues and research findings.

In reviewing the literature, while there are no absolute rules, several guidelines are important. You should think about:

- Is the literature you are looking at peer-reviewed? This should ensure the quality and rigour of the research;
- What research design has been used? Has the author/s made this clear? Have they discussed the research's limitations?
- How current is the material? Where possible you want to ensure that the most contemporary literature is accessed.

4.1   Undertaking Your Review

In undertaking your review, there are a few points you may want to keep in mind. First, when undertaking your literature review be systematic in your approach and ensure that you manage the information you collect in a suitable fashion. Bibliographic programs such as Endnote are a useful tool as they allow citations to be downloaded directly from search engines such as Google Scholar, help manage and categorise the information collected and help keep track of citations as they are added to a document. These types of programs are not necessary (a simple excel file could also work), but they do make the job of managing a literature review easier.

Regardless of the type of information management approach used, there are several pieces of information that must be collected:

- Key search terms used—this avoids duplication.
- Full bibliographic details of each piece of literature read—this can be downloaded directly into programs such as Endnote.
- A list of other associated literature. In Google Scholar, for example, references to any given article will include a hyperlink to related articles and are a good way to expand the list of appropriate material.

You should also make a habit of taking notes as you read, not just highlighting text in an article. Try to summarise the material as you read it. This will help you to be a more active reader and help when it comes to pulling together the material into a coherent review.

As you complete the literature review, keep in mind that a review is not merely a description of work, but should, where possible, provide a critical discussion of the key themes and issues. These themes and issues will be directly related to the research question/s and the key concepts that have been identified.

5   Theory and Research

The methodological approaches and designs used in urban and planning research are not undertaken in a vacuum, devoid of wider intellectual thinking. In fact, the entire research process is generally anchored in a theoretical approach or understanding. A theory is generally understood as a set of propositions that are plausibly related and express the relationship among our different constructs and propositions [13]. Put another way, theories are abstract descriptions of the relationships between concepts that help us understand the world. Importantly, within the whole research context, theories act as a backstory and justification for the research being undertaken and supply the broader framework for understanding and interpreting our research findings.

Many theoretical positions appear abstract in nature and are often referred to as grand theories. These are entrenched in the history and development of a number of social science disciplines and include theories of structural-functionalism, symbolic interactionism, critical theory, post-structuralism and structuration [14]. While these 'grand theories' are important in their own right, they tend to be of less practical use in the research process itself as they are less able to offer a researcher insight as to how they might drive or influence their collection of empirical evidence or its interpretation. In the world of planning research, Lagopoulos [15] has recently argued that monolithic grand narratives are considered too inflexible to be of use to those undertaking planning research or urban analysis. Instead, it is a range of middle-range theories [16] that are more likely to be the focus of research investigation.

These middle-ground theories are less abstract than their grand theory peers and are very often more focused in their domain and varied in their range of applications. In short, middle-ground theories offer less abstract understandings of social phenomena and therefore afford the researcher greater utility in understanding their research within an intellectually conceptual frame. Thinking specifically in terms of urban analysis and planning research, a researcher interested in public transport patronage might use a theory of planned behaviour (a widely used middle-ground theory) to understand the decisions of a population to use public transport [17]. Or a researcher interested in understanding the distribution of socio-economic status throughout a city might turn to social theories of disadvantage or equality to help them understand and structure their research process. Your choice of theories will be guided by your reading around the broad subject area and the extent to which any particular theoretical concepts or approaches fit your research study.

The way theories aid us in understanding our research and interpreting outcomes are linked to the distinction between deductive theory (theory testing) and inductive theory (theory building).

Deductive theory often represents what is considered the most common association between theory and research. A deductive approach, which is usually, but not always, associated with quantitative approaches, focuses on developing a hypothesis (or hypotheses) that is based on some existing theory, and then developing a research strategy to test the hypothesis. Deductive theory is often portrayed as a series of steps (Fig. 1) representing reasoning from the particular to the general. If a causal relationship appears to be implied by a chosen theory, it may or may not be true in a given research context. A deductive design would therefore test to see if this relationship did in fact hold given a particular research context.

Deductive approaches can be explained by means of hypotheses, which can be derived from the propositions of a given theory. Contained within the theory-driven hypothesis is a range of concepts that the researcher translates into researchable entities, thereby guiding the research design and methodologies undertaken. The endgame is to use the research process to test your hypothetical questions, which may lead you to accept or reject the hypothesis and possible rethink or adjust your theoretical principles. This final stage (Fig. 1) can be seen as containing an element of induction (see below), but it is nevertheless the case that much of the process is in fact deductive.

In contrast to deductive approaches, inductive approaches or inductive reasoning approaches start with observations, and theories are proposed towards the end of the research process as a result of observations (Fig. 1). Inductive research involves the

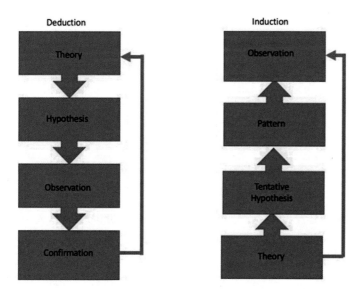

Fig. 1 Deductive versus inductive approaches to the relationship between theory and research

search for patterns from observation and the development of explanations and theories—for those patterns. It is important to note that the inductive approach does not imply simply disregarding theories when formulating research questions and objectives. This approach aims to generate meanings from the collected data and observations in order to recognise patterns and interactions to build a theory. However, the inductive approach does not preclude the researcher from using existing theory to develop the research question to be explored. Moreover, the inductive approach may actually contain a component of deduction in that a researcher may wish to undertake more observation or data collection to further test the conditions in which a theory will be supported or not. This might be called an iterative strategy as it involves weaving back and forth between data and theory.

6   Developing Appropriate Research Design and Methods

Clearly one of the most important aspects of developing a research plan or roadmap is in the choice of appropriate methodological approaches. Armed with research questions and a grasp of the appropriate literature, theories and conceptual approaches, a researcher must decide on the most appropriate design and methodology to use in order to answer their questions. This in turn will be driven by practical considerations (such as level of expertise) and resource considerations (costs versus benefits).

6.1   Research Design

Research design refers to the act of planning how your research will be conducted. It is about getting from point A—a set of research questions or a hypothesis—to point B—an answer to those research questions [18]. Your research design is more than just a decision between taking a quantitative or qualitative approach. Your research design provides guidance into the execution of a particular method and the analytical process that might be followed once the data is collected.

Different research designs are generally associated with different approaches to data collection methods and analysis. Table 1 makes a distinction between a given research strategy (quantitative or qualitative) and a particular research design, although this is not meant to suggest that a researcher can't take a mixed-methods approach combining aspects of different strategies.

Experimental designs, while widely used in many of the science disciplines, are not common in the urban analysis and urban planning research field. Indeed, it would be difficult to undertake an analysis of life in a city (for example) in the same way that we might consider studying the efficacy of a new drug or treatment. Experimental designs involve allocating subjects to two or more groups, experimental groups and control groups to test the efficacy for example of some intervention. Some groups are exposed to the intervention (i.e. a new drug) and others are not (they

Table 1 Research design and research strategy

Research design	Research strategy/approach	
	Quantitative	Qualitative
Experimental	Experimental design employs quantitative comparisons between experimental and control groups with regard to a dependent variable (see Fitzhugh et al. [19], Snellen et al. [20])	Not typically undertaken
Cross-sectional	Survey research or structured observation on a sample at a single time point. Content analysis on a selection of documents (see Elrick-Barr et al. [20, 21], Baum et al. [21, 22])	Qualitative interviews or focus groups at a single time point. Qualitative content analysis on a selection of documents (see Osborne et al. [22, 23], Pearson et al. [23, 24])
Longitudinal	Survey research on a sample at different points in time. Content analysis on a selection of documents related to differing time periods (see Dockery et al. [24, 25], Anastasopoulos Islam (White [8]))	Ethnographic research over a long period of time. Qualitative interviewing on more than one occasion. Qualitative content analysis on document over time (see Newton et al. [25, 26], Breadsell et al. [27, 34])
Case study	Survey research on a single case (see Kamruzzaman et al. [27, 28])	Intensive study by observation or qualitative interviewing (see Connell et al. [29, 35], Maalsen et al. [30, 36])
Comparative	Survey research where there is comparison between two contexts (i.e. cross country) (see Townsley et al. [29, 31], Huang et al. [30, 32])	Ethnographic or qualitative interviewing on two or more cases for the purpose of comparison (see Kunz et al. [31, 33], Sandri et al. [34, 37])

Source Adapted from Bryman [38]

receive a placebo). In this case the researchers would then measure the outcomes for each group and compare the impacts of the intervention on the control groups(s) versus the experimental group(s). The experiments usual occur in highly controlled environments and are referred to as randomised experiments or randomised control trials because a subject's location in a control or experimental group is random.

Quasi-experimental designs are similar in many aspects to experimental designs used in many of the sciences. They are quasi in the sense that they are not as controlled nor are they in many cases randomised. The absence of random assignment to groups often occurs because of the practical difficulties associated with implementing it. A form of quasi-experimental design that has been utilised in urban analysis and planning research is 'natural experiments'. Natural experiments are thought of as experiments as there is still a feel that two groups are being influenced in some way (as in a scientific experiment), but this happens as part of a naturally occurring effort to control a given social setting such as an urban neighbourhood or community. Here it is not possible to randomly assign subjects to experimental and control groups. In the example given in Table 1, Fitzhugh et al. [19] used a quasi-experimental research design with multiple control neighbourhoods to test the level of physical activity of residents. They compared the results from a neighbourhood with a retrofitted urban trail (experimental group) with neighbourhoods with no such trail (control groups), thus making this an experimental design, all-be-it a quasi-one.

More common in urban analysis and urban planning research is the use of cross-sectional designs. A cross-sectional research design entails the collection of data or information from a sample at a single point in time. The data can be quantitative or qualitative in nature and is used to consider associations or patterns between variables of interest in order to answer a research question or hypothesis. Cross-sectional designs are relatively straightforward in terms of process, but they do limit the researcher to drawing conclusions fixed at one point in time, which may limit the types of questions and hypotheses considered.

Researchers using a cross-sectional research design might collect quantitative data themselves (primary data) or obtain data that has already been collected (secondary data), but cross-sectional designs can be also used when a qualitative research strategy is followed. The examples provided in Table 1 cover both quantitative and qualitative strategies. The research by Elrick-Barr et al. [20, 21] presented the result of a project utilising a cross-sectional design that collected primary quantitative data relating to household's responses to climate change. The paper by Baum et al. [21, 22] was again an example of cross-sectional design in that they considered the self-perceived health status of a sample of Australians and tested how this differed according to the types of neighbourhoods people lived in. Rather than collecting this data themselves, the researchers utilised a collection of secondary data sources. The papers by Osborne et al. [22, 23] and Pearson et al. [23,24] are examples of cross-sectional research design where qualitative data has also been used as part of the research strategy. In the case of Osborne et al. [22, 23] the researchers made use of both quantitative and qualitative data to illustrate the different experiences of community between youth and adults as a way to inform urban planners.

In contrast to cross-sectional designs, longitudinal designs expand the research across time allowing the researcher to consider how a particular issue has changed over some period of time. The benefit of this design is that it allows questions and analysis to be framed in terms of understanding how a sample of respondents have been impacted by some issue over time. For example, we might want to know people's satisfaction with their housing changes over time in relation to their changing social and economic situations. A longitudinal design would allow us to do this.

In the case of longitudinal design, a sample is surveyed at a point in time and then surveyed again at a later date. These waves can be completed as many times as the researcher wants, although there is obvious time and financial costs associated with this type of research design. Some of the time and cost issues can be overcome by utilising secondary data, as was the case with the two quantitate examples given in Table 1. The research by Dockery [24, 25] used data from The Longitudinal Study of Australian Children and The Longitudinal Study of Indigenous Children, both secondary data sources, to analyse childhood outcomes and wellbeing, and the association with housing circumstances and understand how these have changed over time. Longitudinal designs are not restricted to quantitative strategies, as the example by Newton et al. [25, 26] illustrates. In this case the researchers undertook a two-and-a-half-year ethnographic qualitative study of 47 long-stay psychiatric inpatients who were discharged into the community.

The idea behind a case study designs is that the researcher undertakes a detailed and intensive analysis of a single case. While a case study approach may seem simple and straightforward it can still often involve the collection and analysis of large amounts of data. A case study approach is often used when the researcher wants an in-depth, multifaceted analysis of complex issues in their real-life settings [26, 35]. The research by Kamruzzaman et al. [27, 28] into transit-oriented developments is an example of the application of quantitative strategies for a case study of the application of TODs in Brisbane. The paper into smart city development by Maalsen, Burgoyne, and Tomitsch [30] utilised findings from their qualitative research to understand aspects of smart city development across a number of case study cities.

The final research design, comparative research, is the act of comparing two or more things with a view to discovering something about one or all of the things being compared [28, 36]. While there are certainly methods that are utilised far more often than others in comparative studies, the use of quantitative analysis is much more widespread [28, 36]. The research noted in Table 1 by Townsley et al. [29, 31] used spatial analysis of secondary data to compare and test a theory of burglar target selection across regions in the Netherlands, United Kingdom and Australia. Similarly, the paper by [30] Huang [32] used cross-country quantitative data and remote sensing to undertake a comparative analysis of the difference in urban form. Although quantitative strategies are most often used, there are examples of comparative studies where qualitative data has been utilised. As an example, the paper by Kunz et al. [31, 33] utilised qualitative data to compare water recycling practices between water utilities in New South Wales.

7 The Practicalities: Bringing It All Together

Once you have developed your research question, undertaken a preliminary review of the literature and thought about the most appropriate approach to take, it is often necessary to draw all these components together into some form of research proposal. Whether you are undertaking an undergraduate research thesis, a Ph.D. or applying for a research grant, it is necessary to put your ideas together in a logical format. This not only helps you to get a grasp of what it is you want to do but is also often a necessity for a university research project or funding body.

There are many ways to put together a research proposal and many organisations have specified formats that should be followed [37, 38]. However, it is usual to include these issues:

- What is your broad research topic or what are the issues you want to address?
- Why is the topic important? Have you identified an important research gap? Are there unanswered questions in a particular area?

- Flowing from this, what is your research question or research questions?
- What research strategies and approaches are you going to use? How do these approaches link back to answering your research questions?
- Why are the approaches you are going to use the most appropriate?
- Who or what will be your research subjects?
- Do you require ethics clearance?
- What resources will you need to successfully complete your project? If appropriate resources are not available, you might need to rethink your approach.
- What is the timetable for your research?
- How will you analyse your data?

Developing a proposal, while not necessarily the most exciting part of the research process, is, as you can see, a crucial component. You should try to keep your proposal simple and straightforward. Try to avoid jargon and try not to write it in a complicated manner. The best advice I ever received about writing a proposal was to write it in such a way that if you got hit by a bus, someone could pick up from where you left off. Finally, when developing your proposal keep in mind that it will never be set in stone; even once you begin the research process there may be times that your proposed approach will need rejigging because of un-foreseen problem.

8 Conclusions

In order to begin undertaking a research project, a researcher needs two important components—a research idea and a research question. Research ideas can come from many sources. Sometimes you are given a research topic to develop or you may decide on a particular topic based on your interests, a class you have taken or a book or journal article you have looked at. Developing a strong and clear research question is often one of the most important parts of the research process as it guides the remaining of the planning and the actual research phases. When undertaking your research, you should keep your research question 'front-of-mind' at all times as this helps avoid going off on a tangent or wasting time on material that will not help you answer your primary questions.

Another important part of the early stages of the research is to undertake a review of the appropriate existing literature. This helps you understand your research area and provides guidance regarding appropriate methodologies and approaches. Literature can include peer-reviewed journal articles, books, reports, conference proceedings and some internet sites.

The development of a strong research question and a good literature review will help the researcher write a strong research proposal. A research proposal usually includes important details about the proposed research project including the aims and background of the project, the significance of the project, the methodological approaches to be used, a timeline and a budget.

Key Points

- Developing a strong research question and a good research design is a crucial part of planning for research
- Your ideas for a research question or topic will come from many places. Very often you will be given a research topic by your lecturer or project leader or, in the case of a consultancy, your client, but you may also be called upon to develop your own research topic. As a student, you might decide on a broad area of research that you looked at in a class or course you have already taken, or your broad topic may come about from an interest you have outside of formal classes.
- Undertaking a concept mapping exercise is often a good way of developing your ideas and cementing your research question (s) and approach.
- Research is not undertaken in a vacuum but is informed by theoretical frameworks.
- Good research design ensures you can move your research in the right direction and not get too side-tracked.

Further Information

For those wanting to get further information about planning your research project and project design, see:

- Martyn Denscombe (2009). Ground rules for social research: Guidelines for good practice. McGraw-Hill Education (UK).
- Elisabete Silva, Patsy Healey, Neil Harris and Pieter Van den Broeck (Eds.). (2014). The Routledge handbook of planning research methods. Routledge.
- Alan Bryman, (2016) Social research methods. 2016: Oxford university press.

References

1. Theodorson GA (1982) Urban patterns: studies in human ecology, Penn State University Press, Pennsylvania
2. Burgess EW, McKenzie RD, Wirth L (1925) The city, University of Chicago Press, Chicago
3. Booth C (1903) Life and labour of the people in London, vol 8, Macmillan and Company, New York
4. Arthurson K (2012) Social mix, reputation and stigma: exploring residents' perspectives of neighbourhood effects. Neighbourhood effects research: new perspectives. Springer, Berlin, pp 101–119
5. Baum S, Mitchell WF (2009) Red alert suburbs: an employment vulnerability index for Australia's major urban regions. University of Newcastle Centre of Full Employment and Equity, Callaghan
6. Wheeldon J, Ahlberg MK (2011) Visualizing social science research: maps, methods, & meaning, Sage, Thousand Oaks
7. Denscombe M (2009) Ground rules for social research: guidelines for good practice, McGraw-Hill Education, Maidenhead
8. White P (2017) Developing research questions, Macmillan International Higher Education, London

9. Bouma GD, Ling R (2004) The research process. Oxford University Press, Oxford
10. Stilwell F (1989) Structural change and spatial equity in Sydney. Urban Policy Res 7(1):3–14
11. Baum S, Hassan R (1993) Economic restructuring and spatial equity: a case study of Adelaide. Aust N Z J Sociol 29(2):151–172
12. Cronin P, Ryan F, Coughlan M (2008) Undertaking a literature review: a step-by-step approach. Br J Nurs 17(1):38–43
13. Kerlinger FN (1996) Foundations of behavioral research
14. Craib I (2015) Modern social theory, Routledge, London
15. Lagopoulos AP (2018) Land-use planning methodology and middle-ground planning theories. Urban Sci 2(3):93
16. Merton RK (1967) On theoretical sociology: five essays, old and new
17. Ambak K, et al (2016) Behavioral intention to use public transport based on theory of planned behavior. In: MATEC Web of Conferences. EDP Sciences, Les Ulis
18. Farthing S (2015) Research design in urban planning: a student's guide, Sage, Thousand Oaks
19. Fitzhugh EC, Bassett DR Jr, Evans MF (2010) Urban trails and physical activity: a natural experiment. Am J Prev Med 39(3):259–262
20. Elrick-Barr CE et al (2016) How are coastal households responding to climate change? Environ Sci Policy 63:177–186
21. Baum S, Kendall E, Parekh S (2016) Self-assessed health status and neighborhood context. J Prev Interv Community 44(4):283–295
22. Osborne C et al (2017) The unheard voices of youth in urban planning: using social capital as a theoretical lens in Sunshine Coast, Australia. Children's Geogr 15(3):349–361
23. Pearson LJ et al (2010) Sustainable land use scenario framework: framework and outcomes from peri-urban South-East Queensland, Australia. Landsc Urban Plan 96(2):88–97
24. Dockery AM, et al (2013) Housing and children's development and wellbeing: evidence from Australian data, Australian Housing and Urban Research Institute, Melbourne
25. Newton L et al (2001) Moving out and moving on: some ethnographic observations of deinstitutionalization in an Australian community. Psychiatr Rehabil J 25(2):152
26. Crowe S et al (2011) The case study approach. BMC Med Res Methodol 11(1):1–9
27. Kamruzzaman M et al (2014) Advance transit oriented development typology: case study in Brisbane, Australia. J Transp Geogr 34:54–70
28. Heidenheimer AJ, Heclo H, Adams CT (1990) Comparative public policy: the politics of social choice in America, Europe, and Japan, St. Martin's Press, New York
29. Townsley M et al (2015) Burglar target selection: a cross-national comparison. J Res Crime Delinq 52(1):3–31
30. Huang J, Lu XX, Sellers JM (2007) A global comparative analysis of urban form: applying spatial metrics and remote sensing. Landsc Urban Plan 82(4):184–197
31. Kunz NC et al (2015) Why do some water utilities recycle more than others? A qualitative comparative analysis in New South Wales, Australia. Environ Sci Technol 49(14):8287–8296
32. Snellen D, Borgers A, Timmermans H (2002) Urban form, road network type, and mode choice for frequently conducted activities: a multilevel analysis using quasi-experimental design data. Environ Plan A 34(7):1207–1220
33. Anastasopoulos PC et al (2012) Hazard-based analysis of travel distance in urban environments: longitudinal data approach. J Urban Plan Dev 138(1):53–61
34. Breadsell JK, Byrne JJ, Morrison GM (2019) Household energy and water practices change post-occupancy in an Australian low-carbon development. Sustainability 11(20):5559
35. Connell S et al (1999) If it doesn't directly affect you, you don't think about it': a qualitative study of young people's environmental attitudes in two Australian cities. Environ Educ Res 5(1):95–113
36. Maalsen S, Burgoyne S, Tomitsch M (2018) Smart-innovative cities and the innovation economy: a qualitative analysis of local approaches to delivering smart urbanism in Australia. J Des Bus Soc 4(1):63–82

37. Sandri O, Hayes J, Holdsworth S (2020) Regulating urban development around major accident hazard pipelines: a systems comparison of governance frameworks in Australia and the UK. Environ Syst Decis 40(3):385–402
38. Bryman A (2016) Social research methods, Oxford University Press, Oxford

Using Systematic Quantitative Literature Reviews for Urban Analysis

Catherine Pickering, Malcolm Johnson, and Jason Byrne

Abstract This chapter discusses how students and early career researchers can use systematic quantitative literature reviews (SQLRs) to answer research questions about cities. These SQLRs can enable a greater understanding of complex patterns, processes, and relationships that occur in human settlements. The chapter begins by overviewing SQLRs, how they differ to narrative and meta-analysis reviews, and what are their benefits. We consider the importance of: starting the right way; being careful to specify the research question(s); exploring the interrelationship between concepts that will guide the literature search; and being clear about the keywords that will be of use for the search, as well as the definition of key terms. Next, we discuss the 15 steps of undertaking a SQLR, examining the opportunities, identifying pitfalls to avoid, and providing some strategies that students can employ to make their review successful. Using examples from existing systematic reviews on topics related to urban analysis, we work through the key principles of rigour, comprehensiveness, repeatability, and criteria for inclusion and exclusion. We then discuss how to develop the database and categorise data, before outlining good practices for analysing and visualising findings. We conclude by pointing to emerging directions on how the SQLR method is evolving.

1 Introduction

Although human settlements, including towns and cities, are highly complex social and ecological entities, there is surprisingly no internationally accepted definition of what comprises a city. This could be because industrial and post-industrial cities are relatively recent, having only grown to prominence over the past 200 years or so. In 1800, most of the world's population was rural—roughly 3% of the world's

C. Pickering (✉)
Griffith University, Gold Coast, QLD, Australia
e-mail: c.pickering@griffith.edu.au

M. Johnson · J. Byrne
University of Tasmania, Hobart, TAS, Australia

© Springer Nature Singapore Pte Ltd. 2021
S. Baum (ed.), *Methods in Urban Analysis*, Cities Research Series,
https://doi.org/10.1007/978-981-16-1677-8_3

population inhabited cities. By 1900, it was closer to 15%. By 2007, more than half of the world's population lived in cities, and by the middle of this century that figure will be closer to three quarters [1]. Despite their complexities in economic transactions, material flows, population movements, and environmental impacts, urban areas comprise only around 3% of the earth's surface area [2]. Yet cities play a fundamental role in the global human population's social, political, economic, and environmental lived experience. Cities are also profoundly reconfiguring the biogeochemical systems of our planet, with implications for humans and non-humans alike [3]. It is little wonder then that the literature about cities is burgeoning and is more than any one person could work through in a lifetime of scholarship. To gain a comprehensive understanding of the research being published about cities increasingly requires systematic and rigorous approaches to undertaking literature reviews.

We need to better understand cities to ensure informed decision-making and to counter myths, stereotypes, or assumptions that underpin discrimination and poor management practices. This is a key task of urban analysis. Rapid urbanisation has brought profound changes to how people move from one place to another, obtain shelter, secure food, water, and clothing, and how we interact with each other and the natural world around us. Cities generate about 80% of the world's economic activity [4]. Also living in cities means that urban residents, planners, and government officials alike are grappling with complex socio-ecological challenges such as waste management, hygiene, and sanitation; water, food, and energy provision; affordable housing; and essential service delivery (e.g., parks, healthcare, transportation, sanitation, safety, and security). Increasingly urban analysis also entails grappling with how to best respond to the impacts of climate change which manifest particularly strongly in cities (e.g., urban heat island impacts)—as well as other environmental and social issues such as natural hazards, food water, and energy security, sustainable livelihoods, social inclusion, and effective governance.

Scholars, practitioners, and decision-makers need access to high-quality data and effective ways of analysing and synthesising that data. High-quality research can help planners, land managers, policy makers, and decision makers to improve the lives of both human and non-human urban dwellers. Undertaking good research requires the use of appropriate methods. It is important to know which methods to use, when, and where. Researchers also need to know how to analyse different types of data and to interpret results, to answer their research questions, and this means being able to think critically. This chapter discusses the importance of understanding what is already known about urban environments in order to direct future research at helping to solve problems. In the chapter we discuss a particular kind of research method—the Systematic Quantitative Literature Review (hereafter SQLR). This method can be used to interrogate the peer-reviewed, scholarly literature to identify knowledge gaps but also to identify which methods have been used to answer particular questions, in which places, and using what analytical frameworks. Before discussing the systematic quantitative literature review, the steps involved, its benefits, and its strengths and weaknesses, it is useful to first specify what we mean by the term urban analysis.

1.1   What is Urban Analysis?

Many disciplines such as geography, planning sociology, anthropology, and urban studies, among others, trace their understandings of built environments back to the Chicago School of Sociology, a group of highly influential scholars working at the University of Chicago in the United States, during the early decades of the twentieth century. These scholars, including Ernest Burgess, Louis Wirth, and Robert Park, felt that much of the work on the existing urban environment consisted of 'untested theories, interesting facts, social work and social reform' [5]. They sought to bring rigour and to develop new methodological approaches to better study cities, including their spatial structure, population dynamics, neighbourhood creation, urban subcultures, ways of life, and social norms and conventions.

The Chicago School developed the human ecology approach to urban studies, based on a systematic application of the principles of plant and animal ecology to the study of human communities. Its focus was on the study of cultural groups in the city, and it was underpinned by an evolutionary and determinist approach to understanding cities, which has since been widely critiqued [6]. But the Chicago School scholars laid the foundation for what we now term urban analysis. Urban analysis is the study of phenomena occurring in urban space, using a combination of analytical tools, prediction models, and forecasting to address contemporary urban issues, including population change, land use change, infrastructure provision, environmental change, hazardous waste generation, and economic restructuring— among others [7]. Many disciplines are involved in contemporary urban analysis, including economics, psychology, spatial science, sociology, geography, ecology, architecture, computer science, and health science. Despite this diversity of perspectives, urban analysis has three identifiable foci: (i) the measurement and evaluation of urban areas; (ii) understanding the mechanisms driving urban phenomena; and (iii) predicting urban change [7].

Urban analysis draws upon diverse social, spatial, economic, and environmental data (e.g., census data, traffic counts, mobile phone location data, and vegetation coverage) [8]. It is often directed at resolving policy issues, and is attentive to the fact that cities are both material and representational or symbolic spaces; they are built from bricks, mortar, concrete, glass, etc. (their materiality) and from the imaginations of architects, planners, developers, artists, and residents (their symbolic dimensions). Researchers undertaking urban analysis might consider how socio-spatial variations in the distribution of environmental benefits and harms are (re)produced through asymmetries in a population's power, income, and ethno-racial composition. Or they could assess how urban tree canopy cover reflects variations in household income. Or how automobile-dependence creates jobs-skills mismatches, how gendered political and social institutions create built environments that are unsafe for women and entrench poverty, or how park design influences the behaviour of humans and non-humans. At the heart of these different concerns are questions about phenomena that occur in cities (e.g., traffic congestion, crime, and habitat loss) and how they vary across scales, space, and time. Literature reviews are an important element in undertaking effective urban analysis because they can inform research design, methodological choices, measures, modes of analysis, and interpretation of results.

2   What is a Systematic Literature Review and Why Do One?

Many scholars and student researchers will be familiar with the traditional narrative style literature review. Such reviews are often used in a research thesis and also in book chapters and published papers. They typically involve detailed searches of scholarly databases to find literature related to a particular research topic. However, such reviews tend not to include a methodology section, including providing information about how the review was undertaken, how the literature was found (e.g., keywords used in a search), and/or criteria for inclusion and exclusion of different studies in the review. Moreover, rarely are patterns in the data quantified (as opposed to more generally identifying key themes in the literature that was examined), and such reviews tend to present only a generalised summary of study findings [9, 10]. In other words, narrative reviews can be highly subjective.

The SQLR method is an effective, step-by-step approach for mapping the literature on a chosen topic [11, 12]. In contrast to narrative reviews, the SQLR method combines a systematic (rigorous, comprehensive, repeatable) method for identifying the relevant literature and employs a quantitative methodology for coding the content of the literature, using categories and subcategories. A key strength of this approach is that it enables researchers to clearly identify *who* has undertaken research on a specific topic, *where* the research was done, and *when*, as well as addressing *which questions* were asked by researchers, using *what methods*, *what they found*, and *how they interpreted* their results, together with their policy recommendations and suggestions for future research [11, 12]. Using a systematic approach enables a researcher to 'map' what is known, identify where gaps exist in current knowledge, recognise emerging trends, point to topics that are hotspots of research, and reveal those which have tended to be neglected.

As a result, a systematic approach can show how a focus on specific issues, methods, locations, and questions may be affecting, or even distorting, our understanding of key knowledge. In turn, a systematic approach enables a researcher to ask how we can better balance the need for more research on a particular topic with factors affecting the creation of new knowledge (e.g., budgets, social implications, and politics). A systematic approach can also help to address important equity issues in research. This is because research is often driven by the needs and context of wealthy, and typically English-speaking countries with highly developed research infrastructure, and good funding and resources [13–15]. The needs of other urban communities, such as those in the global south, are often overlooked [16–18]. Systematic reviews can thus illuminate place-based knowledge gaps and can open up new research agendas.

The SQLR also differs from meta-analysis style reviews, which are an increasingly common method used in research areas such as health and education. While both approaches apply the same systematic process to identifying relevant literature, they differ in what type of literature can be included and how it is quantified. Meta-analyses combine results from several existing quantitative studies that have often used experimental approaches and where data in the individual studies were statistically analysed [10]. This type of review takes the results from those studies and combines them by applying statistical methods such as weighted averages across studies, to provide a more rigorous and comprehensive analysis of the results. But such reviews cannot easily incorporate mixed methods studies or results from qualitative research [10].

An issue sometimes raised with the SQLR method is that it uses a 'vote' counting technique, where different studies (votes) are treated in some respects similar to each other. This is a valid issue when comparing it to meta-analysis methods but can also be seen as a strength. The SQLR approach can assess a wider range of studies—including combining results from quantitative, mixed methods, and entirely qualitative studies. It is also possible to address the vote counting issue by weighting studies within an SQLR based on their 'reliability', where there are well-recognised criteria for doing so. In addition, an SQLR can be used to assess which findings are supported by what types of methods, including determining if there are enough suitable studies for undertaking a meta-analysis, if one is desired.

Another consideration with the SQLR is the size of the literature it can address. Where the pool of literature on a particular topic is very small, with around 30 or fewer specific studies on a topic, a narrative thematic-style review is recommended. Similarly, if a very large literature (over 300 studies) is identified after systematically searching and applying selection criteria, the other approaches such as bibliometric/scientometric methods would be more relevant. These types of approaches are very effective at analysing literatures of thousands of studies, using data already available about the publications in online databases such as Scopus and Web of Knowledge [19].

The analogy we often use to highlight how the SQLR differs from narrative reviews and meta-analyses is the infamous quote from the US Secretary for Defence in 2002, Donald Rumsfeld, when briefing the public about 'Weapons of Mass destruction' [20]. Rumsfeld discussed what he called 'known knowns'—the things we know we know, 'known unknowns', the things we know we do not know, but also the 'unknown unknowns', the things we do not know we do not know. Narrative literature reviews tend to encompass known knowns; they often focus on well-known and recognised studies. Meta-analyses, on the other hand, by systematically searching for literature, will document known unknowns, identifying papers that likely would have been missed with a narrative review. But only an SQLR will assess unknown unknowns— identifying gaps in the literature with important implications for our understanding of the topic, of which we were previously unaware [11]—revealing papers that we may have missed with a narrative review. The SQLR approach can also help understand trends over time, helping to pinpoint when a topic first emerged as a new area of research, and enables us to tie these trends to social, political, environmental, or economic drivers.

Some of the first SQLR publications using the Pickering and Byrne method were in urban research, including a study assessing research on community gardens [16]. That study that found major gaps in the literature reveals that most research had been undertaken in the USA and it was nearly entirely focused on the communities, not the gardens. A second early paper examined urban trees, including their benefits, costs, and the assessment methods used [17]. It was found that although there are a wide range of benefits from street trees, there are also important costs, which were less often recognised, and more research was required because although many studies discussed tree benefits and costs, fewer assessed these benefits and costs. Additionally, much of the literature was from temperate regions, particularly in the USA. Since then, a wide range of journals, authors, and topics (Fig. 1) have started to use the SQLR method for urban analysis, reflecting its applicability to diverse topics and its capacity to analyse diverse studies employing quantitative, qualitative, and mixed methods. In the following sections we use examples from many of these studies to illustrate the key points about the method and how it can be applied in urban analysis research.

Fig. 1 World cloud showing the range of urban topics assessed using SQLR method based on titles and journals of a sample of publications

3 The SQLR Method: Steps for How to Apply the Approach

The following section provides both a simplified exploration of the 15 steps of the systematic quantitative literature review methodology as well as an expansion on the necessary strategies required for early career urban analysis researchers to avoid common pitfalls associated with the methodology. Rather than focusing on the specific aspects of each individual step, as was previously specified in two peer-reviewed papers [12, 21], the steps overviewed here have been combined to reflect the various phases of a SQLR (i.e., ideation, searching, coding, analysis, and writing) while maintaining the core concepts of the initial publications. Additionally, based on the experiences of the authors—conducting reviews as part of research teams, answering questions about the method, assessing the increasing body of literature that utilises the method, and teaching the methodology to thousands of students, early career researchers, and experienced researchers—a list of common pitfalls that limit the quality of reviews is included. We also provide a series of strategies that will assist future applications of the methodology, enabling searches to be rigorous, comprehensive, and repeatable. Accompanying the following tables are examples from published systematic reviews on topics related to urban analysis, and from reviews undertaken by our colleagues and students, in order to provide more context for the steps, pitfalls, and strategies. For video explanations of the steps, example databases, and a list of published SQLR papers, see https://www.griffith.edu.au/griffith-sci-ences/school-environment-science/research/systematic-quantitative-literature-review.

3.1 Ideation and Conceptual Models

Before starting a systematic quantitative literature review, some important thinking and problem identification is required—we term this ideation. What is the topic about that you want to explore, that requires you to undertake a search of the scholarly databases? Let's say, for instance, you want to research cities—a search a scholarly database using the search term 'cities' will likely produce millions of results. You'll need to narrow down your topic to limit the pool of potential papers, by being more specific. What aspects of cities do you want to research—transport, poverty, telecommunications, people's emotions, plant diversity? The reality is that you'll probably have two or three intersecting points of interest that you wish to research—for example, 'transport', 'traffic congestion', and 'commuter behaviour' in Indian cities and potential 'solutions' to known problems. Once you have narrowed down a potential topic like this one, the next step is to sketch out how the concepts intersect, to help you further refine the field of research to a manageable set of papers to review. A Venn diagram is a useful way to begin to conceptualise your chosen topic areas and the geographic location(s) of interest. While a search using the terms above will likely still produce up to tens of thousands of potential articles to read, you'll have already narrowed the pool of potential articles substantially—down from a list of millions. We suggest you spend a bit of time experimenting with a potential topic, allocating an hour or so to type in some keywords into a scholarly database to get a feel for the range of potential papers on your topic of interest. This will also help you to formulate research questions that your SQLR is seeking to answer, for example— 'What measures have been found to reduce traffic congestion in Indian cities' and 'How effective are behavioural interventions in encouraging modal shift in transport use in India?'.

3.2   Selecting Keywords to Search and Setting up Your Database

In general, the first three steps (Table 1) of a SQLR occur concurrently, with researchers testing topics through keyword searches and restructuring research questions with each subsequent step. The starting point of a SQLR typically begins by searching various concepts that are most important to your research using a scholarly database. Google Scholar is often the most accessible during this step, with your search process developing over a series of iterations, to reduce results towards

Table 1 Steps, pitfalls, and strategies for the ideation phase

Steps	Pitfalls	Strategies
Step 1 – Define Topic - Relevant within overall field of urban analysis - Appropriate for SQLR rather than other reviews	- Selecting a topic that is too broad (e.g., single words, concepts, or themes) - Selecting a topic that is too narrow (e.g., hyper-localised with field-specific terms) - Reflecting personal or cultural bias based on assumptions of how things 'should be'	+ Test potential topics in databases for relevancy and breadth of available research papers + Ensure topic is original (e.g., not recently reviewed) and personally engaging + Think about what you know and, more importantly, what you do not know
Step 2 – Formulate research questions - Begin with *who, what, when, where, why,* and *how* - Refine questions as you go as part of the process	- Asking questions that cannot be answered through a systematic quantitative literature review - Letting the questions define the research rather than the research determining the questions - Uncertainty around the central variables and their relationship to the defined topic	+ Find 'gold standard' papers that reflect your topic to guide questions based on unknowns + Review questions following each step and/or each phase to contextualise within discipline + Develop a conceptual model to represent the relationships between variables
Step 3 – Identify keywords - Aim to identify as much of the literature as possible - Utilise synonyms and multiple search terms	- Using terms associated with a particular field or exclusively relevant to a geographic area - Combing a string of disjointed terms or a complete sentence based on the defined topic - Failing to consider the scale of the research (i.e., is it more of a local or international topic?)	+ Search for individual keywords and create a list of synonyms based on the literature + Make use of Boolean Operators (AND, OR, or NOT) to include and exclude relevant keywords + Include qualifiers to refine the search to particular scales when appropriate (e.g., 'small cities')

a manageable number for screening. But Google Scholar will also contain 'grey' literature that has not been peer-reviewed, so it will potentially return many more results that other databases such as Scopus. You'll therefore need to be careful about using 'exclusion' criteria.

Peer-reviewed studies are included but grey literature, other literature reviews, and studies in some languages are excluded (more on this later). Think about your choices carefully as they will shape both the quality of your review and the time it takes to complete the review, as well as how useful it will be for other scholars. Boolean search functions (and/or/not) are necessarily used for this step. For example, the following search results in decreasing results that could be managed in the context of a SQLR: 'green roofs' (32,500 results) AND biodiversity (13,200 results) AND 'ecosystem services' (5,530 results) NOT plants (1,200 results). Within this step, it is important to consider the cultural or discipline biases of certain terms/concepts. For example, the term 'two-wheeler vehicles' is almost exclusively used in literature from India, so if research questions seek to find globally relevant literature—say about non non-motorised transport in cities—then the results for the term 'two-wheeler' will be unnecessarily small and will not reveal the true extent of the published research.

There are some research questions that a SQLR is not equipped to answer, in which case another review method would be more suitable. For example, determining whether there is potential bias in the size of cities that have been studied in urban ecology research would require a scientometric search of thousands or tens of thousands of papers, potentially using a program to trawl the databases. While literature reviews do not necessarily need to be based on a central question or set of questions [22], developing research questions at the outset of your research, and stating them explicitly in any paper or manuscript that you are writing, will ensure the review is in focus [23]. Additionally, through the creation of a conceptual model (such as the Venn diagram discussed above) before progressing into more complex forms [24], you'll be able to better explain the relationships between the key variables that form part of your research topic. An easy step for early career researchers is to settle on a few 'gold standard' papers (i.e., the papers that *must* appear in the database) and 'mine' them for ideas concerning their research questions and their conceptual models—this will help you to model or emulate a proven approach.

The first two steps will inevitably result in a set of keywords that will become the focus of your search. As you'll note from the online videos and published papers on the SQLR method, these keywords will also frame up your personalised database that you'll be using in subsequent steps of the review (e.g., an Excel spreadsheet). In our example above you'd have columns in your database for author, date of publication, field of research, journal name, journal disciplinary area, transport mode, measure effectiveness, city name, etc. You should also carefully define your keywords. We opened this chapter with a definition of urban analysis to make sure that authors and readers had a shared understanding of the topic of this chapter. To develop an effective SQLR, we recommend that you do the same thing. Including the definitions of keywords within the review is recommended because a lack of definitions is often revealed during a review process. Mapping and defining keywords that you use in your search can reduce bias, but it will also ensure that when you are assessing papers you review them the same way as the author intended—are they really talking about values, or are they investigating attitudes, behaviours, beliefs, preferences, or another construct. You might also find that a keyword (e.g., city) has a meaning that is taken for granted but has been poorly defined in the literature. If you carefully define the term with reference to the literature your work will be more robust—and potentially citable by other people who read your work. Generating a list of similar terms/concepts and including them in the search with the 'AND' Boolean Operator may increase the results significantly but will ensure your review targets the right body of literature [25]. Last, before moving on to the next phase, apply qualifiers to search terms (such as city size—mid-size city, or location—India); this can help return a manageable set of results, reducing screening time, and focusing your review.

3.3   Searching Databases and Entering Papers

The most commonly used scholarly databases (see Table 2) in many fields of research—including urban analysis research—are Google Scholar, Scopus, Web of Science, Science Direct, ProQuest Central, GeoBase, and Web of Knowledge. In a review of SQLR papers on urban analysis, we have found that researchers have

Table 2 Steps, pitfalls, and strategies for the searching phase

Steps	Pitfalls	Strategies
Step 4 – Identify & search databases - Decide on a list of appropriate databases - Polish search terms based on database functions	- Utilising a single database for the entirety of the research (i.e., only searching Google Scholar) - Being unfamiliar with the functions of various databases (e.g., 'all fields' versus 'abstract') - Failing to refine keywords following the initial search queries on each database	+ Work closely with supervisors, professors, and librarians to determine appropriate databases + Use advance search features for databases in which one has access (i.e., through university) + Test a series of search terms in each database to decide on an applicable combination
Step 5 – Read & assess publications - Detail and document replicable inclusion criteria - Take extensive notes on exclusion reasoning	- Developing vague inclusion criteria and allowing for exceptions when appropriate - Screening so vast a number of papers that a manageable database seems improbable - Reading unrelated papers rather than screening titles and abstracts for relevancy	+ Document any and all inclusion and exclusion choices to ensure reproducible review methods + Review initial search criteria and determine what is 'doable' within the research timeframe + Develop a systematic approach to review papers (i.e., abstracts, find command, etc.)
Step 6 – Structure database - Develop database with defining categories for data - Create both broad and tight criteria based on values	- Including only a limited set of categories that marginally captures paper information - Focusing primarily on what is known in papers rather than on what is unknown - Including categories that are irrelevant to the research at hand or only reflect one discipline	+ Begin with 'define, describe, & demonstrate' before developing complex multi-tiered categories + Thoroughly review the most relevant papers to determine gaps in the literature/research + Reflect on category language to ensure the majority of papers will eventually meet criteria

tended to use three databases; however, many used only one database (see pitfalls in Table 2); fewer used more than five databases. If you are planning on conducting reviews that are not focused on English-language papers, using different databases will be necessary (see [18] for an example). Once a set of databases are selected, it is important to review their search tips and advance search FAQs, where available. A typical SQLR involves more than six search terms, with some cases including dozens of terms, particularly with the effective use of Boolean operators [25–27]. However, be aware that different databases often require specific nomenclature for searches and will process results differently. For example, Google Scholar replaces 'NOT' with '-' when eliminating terms from results; Scopus allows results to return only from terms contained within titles and abstracts versus Google Scholar's full text searches.

In order to meet the requirements of a SQLR to be systematically replicable, well-defined inclusion and exclusion criteria are essential. As papers are read and assessed during Step 5, screening out grey literature, papers published that are beyond of the scope of the research questions, or papers that don't match a particular criteria will reduce search results from many hundreds papers to around 120, which we have found is the average number of papers for an urban analysis SQLR. Most methods sections in SQLR reviews will include a specific list of inclusion and exclusion criteria [28, 29]. As with many steps of the SQLR process, it is important to comprehensively document any choices you have made as part of the review, such as changing the publication years of your search (if you are searching only within a specific time period) or excluding papers from a particular discipline. Here is it important that you document which papers you have kept in your review and which you have excluded. You can use a PRISMA diagram (http://www.prisma-statement.org/) to show how you have refined the papers at each step. Additionally, while the first few papers should be read in detail to help refine the database, for later stages of the review you can employ the 'find command' (i.e., CONTROL + F) to ensure papers meet the inclusion criteria, thus reducing the time required for publication assessment and data entry into your personalised database.

To further enhance the analytical strength of your SQLR database beyond simply describing (e.g., date range of papers published on a topic), for the keywords used in your search you should add columns with the terms 'defined', 'discussed', and 'demonstrated'. As mentioned earlier, there can often be a lack of definition for key concepts within the literature, a known unknown. But you'll also want to document which papers merely *discussed* a concept you are assessing versus those that undertook empirical research (e.g., interviews, surveys, spatial analysis, pit traps) and produced results and generated findings on the topic (*demonstrated* something). Next, focus on the 'gold standard' papers by developing more categories and columns to incorporate as much information from those papers as possible [30]. We recommend developing more sections and subsections in your database (columns and subcolumns) than will likely end up in the final version of the database, to ensure potentially relevant information is not overlooked, the unknown unknowns. You can always 'roll up' or aggregate data but you can't disaggregate it if you didn't collect it in the first place. Categories should range from the more descriptive details about the papers (e.g., location of study, type of methods used, stakeholders involved, etc.), to specifics involving your research questions (e.g., city size, specific results, etc.), and also to categories for information not explicitly captured within the paper (e.g., climatic regions, proximity to water, governance structure, etc.), which may comprise some of the 'unknown unknowns'. Keep in mind that your database will continue to evolve with each paper assessed; early attempts at being comprehensive can reduce having to go back and rescreen everything multiple times. We recommend initially entering sets of 10% of papers at a time to 'bed down' your database, before proceeding to enter the rest of the papers you have found. That way you can iron out any issues early on.

Several challenges with coding literature in SQLR (Table 3) have been identified from workshops with academics using the method and from student questions. First, we again emphasise how it is important to spend time and effort early on working out the categories and subcategories and testing how feasible, reliable, effective, and relevant they are to the research questions and the actual studies. Those new to SQLR are often tempted to use fewer broad categories and words/terms within them; however, we recommend against this for the following reasons. First, expanding the number of categories and subcategories often saves time later. A more granulated coding, particularly where it possible to just use a binary system (e.g., 0/1 or blanks) rather than entering text (e.g., city name) or scoring studies, can allow gaps to be more easily identified and will make later analysis much faster. For example, where 1

Table 3 Steps, pitfalls, and strategies for the coding phase

Steps	Pitfalls	Strategies
Step 7 – Enter first 10% papers - Fully enter the initial papers and further refine database - Detail changes of inclusion criteria in project notes	- Including papers that lack relevance, reliability, quality, and relevance within the review - Entering a homogenous set of papers (e.g. papers with similar methods or from one approach) - Neglecting to document relevant methodological changes and interesting paper findings	+ Follow standard 'information literacy' guidelines and/or refer to supervisor judgments + Chose papers that cover the breadth of potential information to reduce late-entry revisions + Take comprehensive notes both in and beyond the database to aid referencing and write up
Step 8 – Test & revise categories - Aim to increase level of detail for all categories - Pilot potential charts and data results tables	- Sticking too closely to the initial database (i.e. failing to include additional sub- & categories) - Including broad categories that cover a wide range of data (e.g. "risk" vs. list of multiple risks) - Failing to produce a few charts and tables to test the value of the current sub- & categories	+ Utilise each and every paper as a means to expand the entire database to limit rereviewing + Focus on related terms and concepts within each category to cover as much info as possible + Develop a few tables and review the conceptual model to explore potential missing categories
Step 9 – Enter bulk of papers - Enter papers to ensure database is comprehensive - Document category changes and criteria alterations	- Spending too much time assessing and entering each paper into the database - Changing search terms and/or inclusion criteria without reason or neglecting to document - Overlooking the value of documenting literature reviews, grey-papers, and unused papers	+ Make use of the "find command" to determine presence/absence for categories + Maintain strict inclusion criteria *or* thoroughly document any changes to explain in methods + If papers are excluded but relevant, ensure notes are taken for potential referencing

or blank is used, instead of writing in 'Mumbai', 'Delhi' into a cell in your database, then pivot tables can be easily used in database programs such as Excel, and the data can be transferred to multivariate programs for more advanced analysis. It also tends to be much easier later to combine or summarise subcategories but dividing them later in the process takes considerable work, including having to reread and recode studies. There will, however, be occasions where it is unavoidable to enter text (e.g., you can't anticipate in advance every city you'll find), in which case some degree of later recoding may be necessary.

It is also important that you are clear about the criteria you are using for categories/subcategories, so others can follow and use them—remember, an important strength of the SQLR approach is that it is replicable. Even for simple categories such as the location where the study was conducted can be challenging—study authors may not have been clear about the city, province, or even country name, and there can be political challenges around territories and names of countries, for example, Taiwan/People's Republic of China. Where possible, it is good practice to ensure that if different people are coding papers, a consistent and documented set of procedures are used to ensure that the same results are obtained for all—or a subsample of papers. This increases the consistency, reliability, and repeatability of the review. Also, we recommend recording details about specific coding decisions for future reference, as this will help later if there are queries from reviewers or others about how specific 'hard to code studies' or examples were entered into your database. For example, choosing to code cases with multiple papers rather than individual publications within the database must be explicitly detailed in the methods section [22]. If including papers in the database that were not found exclusively from the search terms, such as through a snowballing method [30] or based on references within publications, those details also need to be stated outright to ensure replicability.

In most cases, grey literature will be excluded from coding and analysis because they are typically not peer reviewed and their results cannot be treated as having the same level of rigor as a peer-reviewed study. However, there are occasions when grey literature can be included through the application of additional inclusion criteria [31] as supplemental information for large research projects with multiple associated publications [32], or in a separate section of the database, to provide additional analysis based on the relevance of the literature (e.g., a policy analysis). As is the case with the main publications screened for coding, the grey literature also needs to be searched efficiently through the use of the 'find command' and keywords, particularly when assessing 100-page reports or theses/dissertations. Lastly, during the coding phase inexperienced researches can be tempted to change their search terms to find papers that better match their initial research questions. While this change can result in more papers, it needs to be documented methodologically and justified in the methods section—and may require reassessing earlier papers.

Table 4 Steps, pitfalls, and strategies for the analysis phase

Steps	Pitfalls	Strategies
Step 10 – Produce & review summary tables - Create tables that list percentages of papers - Highlight any significance or outliers in results	- Focusing exclusively on the total number of papers or percentages within database - Including only vague results in summary tables (e.g., 'most papers utilised case studies') - Inadequately reviewing errors in data entry and/or issues with category definitions	+Begin with brief summaries of each subcategory then produce multivariable/pivot tables +Incorporate citations to relevant papers as well as noting views on outlier results within database +Carefully examine any irregularities within the database and/or restructure categories
Step 11 – Draft methods - Review notes taken during entire research process - Ensure methods are described for replicability	- Neglecting to define keywords, explain categories, or outline inclusion criteria fully - Leaving out the section the search strategy, study selection, or data extraction methods - Including a PRISMA flowchart that does not align with the actual search and entry results	+Utilise notes taken in previous steps to ensure accuracy of the SQLR methodology +Review previous SQLR publications to further inform appropriate methods subsections +Ensure adapted PRISMA is both accurate as well as includes *all* inclusions and additional papers
Step 12 – Evaluate key results & conclusions - Aim for both breadth and depth in topic literature - Make multivariate charts to demonstrate relationships	- Exploring results within single categories or similar variables (e.g., methods versus methods) - Utilising a single chart type (e.g., bar graphs) to demonstrate various findings in data - Focusing on the entire results from summary tables rather than the key insights and findings	+Delve into the connections between different categories and seemingly unrelated variables +Test different chart types with each variable to focus on the emergent properties in data +Choose a few key results that reflect/revise research questions to focus on in the conclusion

3.4 Analysing Your Data and Presenting Results

Many SQLRs have presented results from the analysis phases (Table 4) as tables showing the number and/or percentage of studies for the different categories of data. However, increasingly more sophisticated and effective ways of summarising and displaying the data are available. With the use of pivot table functions in programs such as Excel and graphics, options data of counts or percentage of studies can be displayed across variables more effectively as graphics such as spider diagrams [33] or tree maps [34]. Graphics showing one variable against another are also being used to show relationships in the data, including staked columns or bars of number or percentage of studies by categories [17, 34, 35], as well as temporal patterns in the number of studies [18, 25, 36, 37], journals publishing studies [23], location of studies [32], methods used [34, 38], or themes in studies [30, 38].

Many SQLRs include geographical heat maps showing the numbers and/or percentage of studies per country or city [15–18, 23, 25, 26, 34, 36, 39, 40], but also geographic themes within studies Ulpiani [41]. Some have used variations on Venn diagrams to show patterns in the overlap among themes within studies [40, 42], or semi-graphic tables to compare types of research versus scale [40]. Other two-way comparisons of categories include comparing the latitude and population size of cities in studies and comparing perception and valuation variables with biodiversity scales [36]. Some studies have used statistics (Chi-squared) test to identify significant changes in topics covered by studies over time [43]. Multivariate analysis has also been used to compare multiple themes and hence identifying more complex patterns among studies [15, 44, 45]. And some SQLRs are now mapping data spatially, for instance generating world maps that show the distribution of studies by city, country, or continent.

Some reviews have also combined the SQLR methodology with Leximancer thematic analysis of the studies found in the review (see Fig. 2), where pdf files of every paper being analysed are entered into Leximancer to undertake a thematic and relational analysis of text in those documents. Not only does this add richness to the review, it can also enhance concept specification [45, 46]. Others have added a bibliometric methodology, including citation and co-citation analysis to their SQLR reviews [23]. Such increasing sophistication in the analysis of SQLR is consistent with the maturation of a methodology and more easily allows emergent properties in the literature(s) to be identified. All advanced analysis approaches should be detailed in the methods section, in addition to the search approaches documented throughout the initial three phases. Additionally, it is common for papers employing SQLR

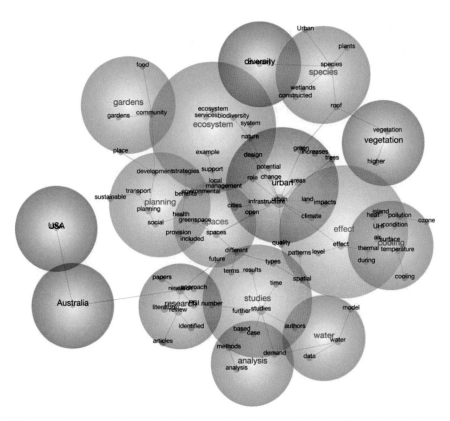

Fig. 2 Leximancer analysis showing the range of topics assessed using SQLR method for urban analysis reviews

methods to include adapted PRISMA diagrams [23, 25, 26, 37, 47, 48] either as part of the methods section or results section of the final manuscript. While some urban analysis SQLR papers have condensed methods sections, we recommend including a comprehensive step-by-step examination of the methodological decisions made throughout the review [23, 28, 30–32, 36]. Alternatively, including supplemental material or appendices, of either methods or the included papers, can help ensure papers meet publication length and figure and table specifications [18, 24, 27, 32, 37].

When drafting the results and discussion sections during the writing phase (Table 5), it is important to highlight the gaps in the literature and the patterns that arise from analysis. In general, the gaps and patterns will answer your research questions, determined in Step 2 or revised based on unexpected results from the analysis. One of the most common patterns found in literature reviews is that most research is written in English, conducted by research institutions in North America and Europe, particularly in the USA, and that most studies are in urban areas in these continents, and focused upon a select number of cities. These geographical hot and cold spots are very important in terms of the influences on not only what we know, but also what we do not know. The research reported in many urban analysis studies not only directly address concepts such as urban social, political, and environmental justice, but often does not represent the lived reality of most of the world's urban populations. Nearly a billion people now live in informal settlements (sometimes termed slums),

Table 5 Steps, pitfalls, and strategies for the writing phase

Steps	Pitfalls	Strategies
Step 13 – Draft results & discussion - Write key findings as the discussion section - Connect results included to the key findings	- Concentrating on the totals within categories rather than the patterns within the data - Including results from every category and subcategory (i.e. long list of percentages/totals) - Failing to answer the research questions sufficiently or altering questions to reflect findings	+ Refer to the supply/demands in literature as well as other pattern-based results analysis + Highlight key findings in discussion section and include only the most relevant results + Review research questions and either refer to them in conclusion or modify them to align
Step 14 – Draft intro, abstract, & references - Focus on Steps 1-3 for introduction section - Ensure accurate referencing of cited literature	- Presenting a buffet of literature without a "golden thread" from which your paper is focused on - Showcasing *only* what is already known in the literature (i.e. mimicking previous reviews) - Writing the abstract without utilizing key results or research question implications	+ Lay out a carefully stepped argument that leads to your aims and ties the paper together + Focus on the unknowns or gaps in the literature to highlight the significance of the SQLR + Summarize the key findings in the abstract with percentages based on data analysis
Step 15 – Revise paper till ready for submission - Review manuscript publication guidelines - Reread and revise paper until submission	- Assuming the first draft of the paper both includes all the results and is clearly delivered - Employing the techniques on the SQLR method inadequately (i.e. not following strategies) - Including expansive methods, results, or literature lists in text detracting from word count	+ Rewrite the paper through a series of drafts to ensure all the information is convey aptly + Review literature about SQLR method to ensure manuscript is ready for publication (see link) + Consider the use of appendixes where appropriate (i.e. including a modified version of database)

but there are no SQLR papers that we are aware of that have examined the literature on informal settlements.

In some SQLRs the disconnect between where research has been conducted and the demand/need for it has been quantified by calculating number of studies versus population, area, or diversity measures [15, 49]. These types of analyses starkly demonstrate how the geographic dominance of the global north shapes perceptions about cities and urban space and point to the need for research from the global south. Some SQLRs also discuss factors affecting the supply of research [15, 16], including how there are more researchers, more funding, and more research infrastructure in the USA and some parts of Europe. Moreover, English language publications dominate knowledge production. They illuminate how cultural and economic factors can drive higher citation rates, increase the prominence of some journals over others, and create social biases against some types of research and some regions. Multi-lingual approaches to SQLR are attempting to address some of these biases and lacunae, providing access to findings published in languages other than English [14, 18]. We would encourage scholars, wherever possible, to continue efforts to overcome privileged forms of knowledge-production.

4   Conclusions

In this chapter we have discussed a method for undertaking systematic quantitative literature reviews—both as a type of urban analysis and as a technique for allowing a better understanding of papers that have been published on specific topics about urban analysis. We have sought to provide insights, recommendations, suggestions, as well as ideas for avoiding the usual pitfalls some people have encountered when undertaking a SQLR. We have overviewed the process rather than providing a detailed description of each step (see [11, 21] for a step-by-step discussion). As we have noted, the main advantage of the SQLR approach is to support researchers in systematically analysing literature on a particular topic and identifying knowledge gaps, as well as patterns in what is known and how it is known. Early career researchers, or those unfamiliar with the rigor of academia, often struggle with developing their initial publications and can become overwhelmed by the task of analysing a large number of papers. The approach described here offers an alternative to narrative reviews methods, and/or can provide the basis for future meta-analyses. By enabling scholars to carefully and meticulously record their decisions in which literature they have searched and how they have searched, as well as employing a range of analytical techniques and visualisations, the systematic quantitative literature review method has become an accessible tool helping students and early career researchers to publish early and often, while open up new research agendas earlier in their careers than might otherwise be possible.

The SQLR, as demonstrated by the number of papers that are employing the approach, is a valuable and effective method that complements, but differs from, other methods such as narrative and meta-analyses. The approach is particularly beneficial to those who are new to literature reviews, as well as to those new to a research topic, as it uses a structured and replicable process. The results of SQLR are often publishable, and published papers using this method tend to be highly cited because they provide important insights, illuminate knowledge gaps, and chart future research directions. The SQLR process itself is clearly defined through the 15 steps ensuring papers achieve a significant level of rigour. At the same time, there is also sufficient flexibility within the methodology to ensure it meets the diverse needs and intentions of researchers across the academy, from the global north and south, and within policy and practice.

Emerging trends include the adaptation of the method for reviewing projects (e.g., types of property development), case law, government policies and grey literature. There are also new types of analyses, including the adaptation of tools from psychology to quantify the use of different types of research methods employed in studies. And new types of data analysis and visualisation of results are also emerging (e.g., Leximancer analysis) allowing researchers to synthesise findings from increasingly sophisticated analytical techniques. As long as each step of the SQLR process is followed, and the decisions made are detailed in full, a systematic quantitative literature review can meet diverse research goals—extending the review in exciting new directions that were not even considered possible just a decade or so ago. For example, we can envisage teams of researchers from across the globe systematically analysing research on a topic published in a wide variety of languages to identify blind spots that scholars did know existed. If the technique is adapted for machine learning, there is also the promise of finally achieving a comprehensive understanding of complex and rapidly evolving research topics such as urban climate change adaptation. There is much to learn about cities and this method is a useful way to begin.

Key Points

- Urban analysis refers to the measurement and evaluation of urban areas; understanding the mechanisms driving urban phenomena; and predicting urban change.
- Scholars, practitioners, and decision-makers need access to high-quality data and effective ways of analysing and synthesising that data, including peer-reviewed studies.
- The systematic quantitative literature review (SQLR) is a method for undertaking literature reviews, based on the key principles of rigour, comprehensiveness, and repeatability.
- The method has 15 steps that begin with ideation and conceptualisation, proceeding to searching databases, coding papers, analysing patterns and trends, and then writing up the findings.
- SQLR reviews enable emerging and established researchers to rapidly assess a diverse body of research on a topic, that uses different methods, to identify the known knowns, known unknowns, and unknown unknowns in research with accuracy and credibility.

Further information

For those wanting to get further information about using systematic quantitative literature reviews, there are a number of useful guides.

Additional resources including videos, example studies, power points, etc. are available at https://www.griffith.edu.au/griffith-sciences/school-enviro nment-science/research/systematic-quantitative-literature-review.

- Pickering C, Byrne J. How to find the knowns and unknowns in any research. The Conversation [Internet]. 2014 September 24, 2020. Available from: https://theconversation.com/how-to-findthe-knowns-and-unknowns-in-any-research-26338.
- Pickering C, Byrne J. The benefits of publishing systematic quantitative literature reviews for PhD candidates and other early-career researchers. Higher Education Research & Development. 2014;33(3):534–48.
- Pickering C, Grignon J, Steven R, Guitart D, Byrne J. Publishing not perishing: how research students transition from novice to knowledgeable using systematic quantitative literature reviews. Studies in Higher Education. 2015;40(10):1756–69.

References

1. Roberts L (2011) 9 Billion? Science 33(6042):540–543
2. Liu Z, He C, Zhou Y, Wu J (2014) How much of the world's land has been urbanized, really? a hierarchical framework for avoiding confusion. Landscape Ecol 29(5):763–771
3. Satterthwaite D, Dodman D (2013) Towards resilience and transformation for cities within a finite planet. SAGE Publications: London, England
4. Balland P-A, Jara-Figueroa C, Petralia SG, Steijn MP, Rigby DL, Hidalgo CA (2020) Complex economic activities concentrate in large cities. Nature Human Behav 4(3):248–254
5. Palen JJ (1975) The urban world. McGraw-Hill Companies: New York, USA
6. Dear M (2002) From Chicago to LA: making sense of urban theory. SAGE Publications: London, England
7. Asami Y, Sadahiro Y, Ishikawa T (2009) New frontiers in urban analysis. CRC Press, Bacon Rota, USA
8. Ratti C, Frenchman D, Pulselli RM, Williams S (2006) Mobile landscapes: using location data from cell phones for urban analysis. Environ Plan 33(5):727–748
9. Boote DN, Beile P (2005) Scholars before researchers: On the centrality of the dissertation literature review in research preparation. Educ Research 34(6):3–15
10. Petticrew M, Roberts H (2006) Systematic reviews in the social sciences: a practical guide. Blackwell Publishing, Oxford, USA
11. Pickering C, Byrne J (2014) How to find the knowns and unknowns in any research. The Conversation [Internet]. Accessed September 24, 2020. Available from: https://theconversation.com/howto-find-the-knowns-and-unknowns-in-any-research-26338
12. Pickering C, Byrne J (2014) The benefits of publishing systematic quantitative literature reviews for PhD candidates and other early-career researchers. Higher Educ Res Dev 33(3):534–548
13. Ballantyne M, Pickering CM (2015) The impacts of trail infrastructure on vegetation and soils: current literature and future directions. J Environ Manage 164:53–64

14. Barros A, Monz C, Pickering C (2015) Is tourism damaging ecosystems in the Andes? Current knowledge and an agenda for future research. Ambio 44(2):82–98
15. Pickering C, Rossi SD, Hernando A, Barros A (2018) Current knowledge and future research directions for the monitoring and management of visitors in recreational and protected areas. J Outdoor Recreation Tourism 21:10–18
16. Guitart D, Pickering C, Byrne J (2012) Past results and future directions in urban community gardens research. Urban Forestry Urban Greening 11(4):364–373
17. Roy S, Byrne J, Pickering C (2012) A systematic quantitative review of urban tree benefits, costs, and assessment methods across cities in different climatic zones. Urban Forestry Urban Greening 11(4):351–363
18. Rupprecht CD, Byrne JA, Garden JG, Hero J-M (2015) Informal urban green space: A trilingual systematic review of its role for biodiversity and trends in the literature. Urban Forestry Urban Greening 14(4):883–908
19. Verrall B, Pickering CM (2020) Alpine vegetation in the context of climate change: a global review of past research and future directions. Sci Total Environ, 141344
20. Logan DC (2009) Known knowns, known unknowns, unknown unknowns and the propagation of scientific enquiry. J Exp Bot 60(3):712–714
21. Pickering C, Grignon J, Steven R, Guitart D, Byrne J (2015) Publishing not perishing: how research students transition from novice to knowledgeable using systematic quantitative literature reviews. Stud High Educ 40(10):1756–1769
22. Lebeau P, Verlinde S, Macharis C, Van Mierlo J (2017) How can authorities support urban consolidation centres? a review of the accompanying measures. J Urbanism: Int Res Placemaking and Urban Sustain 10(4):468–486
23. Nikulina V, Simon D, Ny H, Baumann H (2019) Context-adapted urban planning for rapid transitioning of personal mobility towards sustainability: a systematic literature review. Sustainability 11(4):1007 (online)
24. Hegetschweiler KT, de Vries S, Arnberger A, Bell S, Brennan M, Siter N et al (2017) Linking demand and supply factors in identifying cultural ecosystem services of urban green infrastructures: A review of European studies. Urban Forestry Urban Greening 21:48–59
25. Parker J, Simpson GD (2018) Public green infrastructure contributes to city livability: a systematic quantitative review. Land 7(4):161
26. Knapp S, Schmauck S, Zehnsdorf A (2019) Biodiversity impact of green roofs and constructed wetlands as progressive eco-technologies in urban areas. Sustainability 11(20):5846
27. Ulpiani G (2019) Water mist spray for outdoor cooling: a systematic review of technologies, methods and impacts. Appl Energy 254:
28. Yu R, Burke M, Raad N (2019) Exploring impact of future flexible working model evolution on urban environment, economy and planning. J Urban Manag 8(3):447–457
29. Köck R, Daniels-Haardt I, Becker K, Mellmann A, Friedrich AW, Mevius D et al (2018) Carbapenem-resistant Enterobacteriaceae in wildlife, food-producing, and companion animals: a systematic review. Clin Microbiol Infect 24(12):1241–1250
30. Rahim MS, Nguyen KA, Stewart RA, Giurco D, Blumenstein M (2020) Machine learning and data analytic techniques in digital water metering: a review. Water 12(1):294
31. Vieira LC, Serrao-Neumann S, Howes M, Mackey B (2018) Unpacking components of sustainable and resilient urban food systems. J Clean Prod 200:318–330
32. Brzoska P, Spāǧe A (2020) From city-to site-dimension: assessing the urban ecosystem services of different types of green infrastructure. Land 9(5):150
33. Herington M, Van de Fliert E, Smart S, Greig C, Lant P (2017) Rural energy planning remains out-of-step with contemporary paradigms of energy access and development. Renew Sustain Energy Rev 67:1412–1419
34. Teles da Mota VT, Pickering C (2020) Using social media to assess nature-based tourism: Current re-search and future trends. J Outdoor Recreation Tourism 30
35. Grogan LF, Ellis W, Jones D, Hero J-M, Kerlin DH, McCallum H (2017) Current trends and future directions in koala chlamydial disease research. Biol Cons 215:179–188

36. Botzat A, Fischer LK, Kowarik I (2016) Unexploited opportunities in understanding liveable and biodiverse cities. A review on urban biodiversity perception and valuation. Global Environ Change 39:220–33

37. Boulton C, Dedekorkut-Howes A, Byrne J (2018) Factors shaping urban greenspace provision: a systematic review of the literature. Landscape Urban Plann 178:82–101

38. Turner JA, Babcock RC, Hovey R, Kendrick GA (2017) Deep thinking: a systematic review of mes ophotic coral ecosystems. J Mar Sci 74(9):2309–2320

39. Dupre K (2019) Trends and gaps in place-making in the context of urban development and tourism. J Place Manag Dev 12:102–120

40. Koc CB, Osmond P, Peters A (2018) Evaluating the cooling effects of green infrastructure: A sys tematic review of methods, indicators and data sources. Sol Energy 166:486–508

41. Ulpiani G (2020) On the linkage between urban heat island and urban pollution island: three-decade literature review towards a conceptual framework. Science of Total Enviro 141727

42. Wortley L, Hero JM, Howes M (2013) Evaluating ecological restoration success: a review of the lit erature. Restor Ecol 21(5):537–543

43. Wraith J, Norman P, Pickering C (2020) Orchid conservation and research: an analysis of gaps and priorities for globally Red Listed species. Ambio, 1–11

44. Milcu AI, Hanspach J, Abson D, Fischer J (2013) Cultural ecosystem services: a literature review and prospects for future research. Ecology Soc 18(3):online

45. Robb L, Lawson C, Pickering C, Bikundo E (2021) Schmitt's life within the academy since 2001. In: Tranter K, Bikundo E (eds) Carl Schmitt and The Buribunks: Technology, Law, Literature. Routledge: London, England In Press

46. Thomas S (2014) Blue carbon: knowledge gaps, critical issues, and novel approaches. Ecol Econ 107:22–38

47. Dedekorkut-Howes A, Torabi E, Howes M (2020) When the tide gets high: a review of adaptive responses to sea level rise and coastal flooding. J Environ Plann Manag, 1–42

48. Parker J, Zingoni de Baro ME. Green infrastructure in the urban environment: A systematic quantitative review. Sustainability. 2019;11(11):3182

49. Steven R, Pickering C, Castley JG (2011) A review of the impacts of nature based recreation on birds. J Environ Manage 92(10):2287–2294

Secondary Data in Urban Analysis

Scott Baum

Abstract Data for use in urban analysis and planning research can come from many sources. One particular useful means of obtaining data is through the use of secondary data sources. This chapter introduces the basics of accessing secondary data, with a focus on Australian-based data. It discusses the advantages and disadvantages of using secondary data in research and presents details of a range of potentially useful secondary data sources available in the Australian context. These include data collected in large-scale social surveys together with data that is collected by governments as part of their daily business, including large-scale census collections. The chapter concludes by introducing the area of big data as a useful and exciting form of secondary data.

1 Introduction

When a researcher considers the types of data they wish to use, they are confronted with two broad choices—they can either collect their data themselves or as part of a team, or they can utilise data that has already been collected. Within the context of urban analysis significant use is made of data that has been collected by another party. Urban areas or cities are often too large or complex for researchers, who often have limited resources, to undertake large-scale surveys or other data collection methods. Because of this, data from secondary sources has formed the basis of many of the classic studies into urban issues and continues to be one of the mainstays in contemporary analysis [1]. The work of the Chicago School relied heavily on secondary data, and the use of secondary data was especially important to the work of Eshref Shevky and his associates [2, 3], who developed the descriptive accounts of residential areas that were labelled 'social area analysis' (see box).

S. Baum (✉)
Griffith University, Brisbane, QLD, Australia
e-mail: s.baum@griffith.edu.au

© Springer Nature Singapore Pte Ltd. 2021
S. Baum (ed.), *Methods in Urban Analysis*, Cities Research Series,
https://doi.org/10.1007/978-981-16-1677-8_4

Social Area Analysis

Social area analysis was originally conceived by Shevky and Williams [3] and later by Shevky and Bell [2] and involved the measurement and analysis of city characteristics disaggregated to small census areas (census tracts). The theory and social concern driving this type of analysis was acknowledgement of the changing nature of cities that were undergoing rapid urbanisation during the post-World War 2 period. For these researchers, the focus was on areas as aggregates of individuals rather than on individuals within areas.

The representations created by social area analysis and associated output were an attempt to reduce the complexities of cities into simplified typologies along three dimensions—economic status, family status and ethnic status. Shevky and his associates undertook their analysis on census tracts (small geographic areas used to collect census data in the USA). Each census tract was given three scores or index numbers, representing the three dimensions. Each of the three dimensions was derived from a number of individual variables. Economic status, for example, was derived from eight variables combined into one number using a simple algorithm (the sum of the eight variables divided by eight to produce a simple unweighted index). The eight individual variables were: years of schooling; class of worker; major occupational group; value of home; plumbing and repairs; persons per room; and heating and refrigeration. Once the three dimensions were calculated census tracts with similar configurations of scores were grouped together into larger units referred to as 'social areas'. Cities could then be understood as a mosaic of social areas. As a conceptual and methodological approach, the work by Shevky and others was replicated in other cities both in the United States and internationally [4–7], but was also heavily criticised [8, 9].

Following the criticism on social area analysis, the approach and ideas were refined into a more general factorial ecology by Berry and Kasarda [10], who followed [11] in developing the 'method par excellence for comparing cross-nationally (and intra-nationally) the ecological differentiation of residential areas in urban and metropolitan communities' [11]. These factorial ecological approaches engaged, for the time, sophisticated statistical approaches including principal components analysis and factor analysis to develop indices similar to the social area analysis approach. These similarities have led some to suggest that while important distinctions can be drawn between social area analysis and factorial ecology, both have as one of their end goals the spatial depiction of social processes or structures within cities and urban areas.

While approaches such as social area analysis and factorial ecology have become less popular, there is in contemporary urban analysis the approach known as geodemographic classifications [12] which aim to produce similar outputs as were developed by these earlier researchers.

In contemporary urban analysis, secondary data continues to be used across a range of research projects. A browse through any of the mainstream urban analysis journals (i.e., Urban Studies, Applied Geography or the Australian-focused journal Urban Policy and Research) illustrates the extent of secondary data use in contemporary analysis including research on residential segregation [13, 14], housing affordability [15, 16], social disadvantage [17, 18], residential satisfaction [19] and social issues including crime [20–22].

In a formal sense, secondary data can be defined as data that has already been collected via primary sources and which is then made available for researchers to use for their own research. Simply, it is a type of data that has already been collected in the past by someone else. With the increasing usage of modern computing and the internet, secondary data from around the world has been opened up for use by researchers. Just sitting at my desk I can access census data from the United States via the United States Census Bureau (www.census.gov/) or data on household spending patterns in England via the United Kingdom Household Longitudinal Survey (www.understandingsociety.ac.uk/). For urban analysts today, there are literally thousands of options for gaining access to secondary data. Secondary data can come in quantitative forms or qualitative forms and can include data from censuses, information collected by government departments, organisational records and data that was originally collected for other research purposes. More recently there has been move towards using 'big data' and crowd sourced data or data from citizen scientists.

2   Why Use Secondary Data?

Many people think that if they have to conduct a research project of some kind, then it is necessary to physically collect the data themselves or at least pay someone to collect the data for them. However, this could not be further from the truth, and as a means of obtaining data for research, secondary data presents significant advantages. Most often the reason we might use secondary data is that it allows the researcher to avoid data collection problems including the costs, both financial and time, that are related to collecting data yourself [23].

Imagine if you were wanting to conduct research about residential satisfaction and you were interested in understanding the difference between the levels of satisfaction between low income and high-income suburbs across Australian cities. To obtain a reasonable sized sample that represented individuals and/or households across different suburbs would in itself be a costly and time-consuming exercise. You would need to obtain a sample, produce a survey instrument, pilot the survey, employ interviewers and enter the data from the survey. If you had unlimited time and all the research money you could need, then this might be a possibility. But in reality, very few of us have these kinds of resources at our disposal. However, these financial and time costs can be significantly reduced by accessing data from sources such as the Household and Income Dynamics Australia (HILDA) survey which contains many of the variables such a research project would require. As an example, the study by

Baum, Arthurson and Rickson [19] into residential satisfaction used the HILDA data to analyse how residential satisfaction differed according to both the characteristics of the individual respondent and the type of residential neighbourhood they lived in. The data they analysed consisted of over 8,000 respondents. By using the data that was already collected the researchers were able to complete their study without having to undertake costly and time-consuming data collection that would have been needed.

Using secondary data also sometimes allows the researcher to undertake analysis overtime to investigate the ways in which a given issue or research question has changed over time. The HILDA survey mentioned above is a longitudinal survey where the survey team go back to the same sample of people each year and ask a similar set of questions. We could therefore use this data to expand the research by Baum, Arthurson and Rickson [19] and look at how residential satisfaction has changed over time. Similarly, data from the Australian Bureau of Statistics is often available across several years. The Census of Population and Housing is conducted every five years and provides a range of demographic, social and economic data at various levels of spatial aggregation. While there are some changes in the spatial units used and how they are aggregated, it is usually a straightforward exercise to analyse data over at least a 10-year period. Here a researcher might, for example, want to look at how the age distribution of a city or region has changed over time or the ways in which patterns of housing consumption have shifted. The paper by Randolph and Tice [17] provides another example, whereby the researchers used census data for 1986 and 2011 to investigate the trends in disadvantage across the main metropolitan cities of Australia.

A third benefit of utilising secondary data is that very often, the data is of a high standard or quality. Large data sets that have been collected by professional survey organisations often have the distinct advantage of being rigorously sampled and collected. Very often in surveys the use of a well-constructed sample drawn from a quality sampling frame means that the resulting data is close to being representative as possible. While professional survey companies are faced with the same problems you or I might face when trying to conduct a survey—for example, respondents not completing their questionnaire (survey non-response)—they are well equipped with appropriate methods and approaches to overcome these problems. Therefore, whereas a student conducting a survey might just have to accept that there will be a large percentage of non-response, professional survey organisations are able to deal with these issues by extensive follow-up procedures resulting in a better representative sample and therefore higher quality data.

Another advantage of using secondary data is that it allows a researcher to undertake comparative research across different countries. The availability of comparative national and sub-national data in the form of official censuses or large household surveys combined with ease of access via online portals means that a researcher could, if they wished, consider how the same research question plays out across different countries. For example, going back to the question of residential satisfaction, the Australian data available from the Household and Income Dynamics Australia (HILDA) survey has equivalent counterparts in the United Kingdom England via the United Kingdom Household Longitudinal Survey (www.understandingsociety. ac.uk/), the United States via the United States Panel Study of Income Dynamics (www.psidonline.isr.umich.edu/) and Europe via The European Community Household Panel (www.ec.europa.eu/euro-stat/web/microdata/european-community-hou sehold-panel). This means that a researcher could compare countries or regions to find out how populations differ in terms of their level of residential satisfaction.

A final advantage of using secondary data relates to the ability to spend more time analysing the data and thinking about the outcomes. This is especially pertinent to a student undertaking a small-term project or an honour's thesis or for a researcher who has to meet a strict timeline as may be the case in some evaluation research. Quite often the actual data analysis associated with a research question can be quite complex and time-consuming. Very often the analysis will take a wrong turn or will need to be modified in some way because of difficulties with the data structure or variable construction. The analysis of data is not a straightforward process, and so having extra time afforded by using readily available data (provided it is suitable) can be a clear benefit to the final product.

3   Disadvantages of Secondary Data

While the use of secondary data can be a significant asset in many research situations, there are also a range of pitfalls that can make its use subject to caution. One of the more significant challenges facing researchers who access secondary data relates to the inability to design the data collection process in a way that allows you to best answer your research question. If we want to measure a particular attribute and concept, the use of secondary data may not always be suitable when the data has been collected for a different purpose. Questions may have been asked in a way that doesn't necessarily match what you want, data items may have been collapsed into smaller categories (i.e. chronological age into age categories or continuous income data into ranges) or the data may simply not match up with the concepts you are attempting to measure [24]. The consequence may be that a researcher using secondary data may not be able to focus in any specific way on the original research question and may have to trade-off data accessibility against the desire to stay with the originally hypothesised research question. An alternative hypothesis may have to be considered, or the researcher may be forced to rethink their research ideas or questions.

For example, let's say you are attempting to undertake an analysis of the ways in which different parts of the city are impacted by rising petrol prices and decide to use census data to measure this. The census doesn't contain any variables relating to consumer habits in relation to rising prices, but it does have data on the number of cars per household, income levels and the use of cars to travel to work. Could you use these indicators to develop a measure? You might be able to identify suburbs with a large number of cars per household, suburbs with low incomes and suburbs where a car is used to travel to work, but would this actually tell you what you want? Sure, you might find out which suburbs have low-income households, more than 1 or

2 cars per household and drive to work, but you will find out nothing about whether the households in these suburbs are actually struggling to meet the cost of fuel. In this situation, some care should be taken in making the leap from what you think these variables might tell you and what they are actually telling you.

Another potential pitfall surrounding the use of secondary data revolves around data quality. Putting aside the comments above regrading data quality, how sure can you be that the data you are going to use has been collected in a robust and ethical manner? While we might expect that data collected through official government sources may be robust, census or statistics organisations often voice concern regarding the level of accuracy in self-completed surveys such as population censuses. There has been wide ranging concern regarding the underreporting of income and salaries [25, 26] as respondents may question the confidentiality of responses given to governments who also operate taxation offices and run data matching. Another less serious example might relate to the sudden upswing in Jedi Knights in Australian society [27], as respondents to the census note their religion as being closely aligned to the Star Wars movie franchise rather than an officially recognised religion.

Another potential disadvantage of using secondary data relates to the lack of familiarity with the data collected and the level of complexity of data sets. In situations where you collect your own data, it is likely that you will be familiar with exactly what is contained in the survey and the types of data collected. This is obviously a benefit when it comes to analysing your data as you know exactly what format the variables are in and how the questions were asked. This is not the case with secondary data and can therefore add an extra level of complexity to the process of analysis. For example, the user manual for the Household and Income Dynamics Australia (HILDA) survey runs to over 200 pages and comprises several different data sets. If a researcher wishes to use this type of data set, then there will necessarily be a period of familiarisation required to understand the structure of the data sets, the types of information contained in the data sets and the ways in which the variables are coded.

4   Sources of Secondary Data for Urban Analysis

The types of secondary data available for use in urban analysis is wide and varying. They can come from government organisations, academic researchers, or increasingly from corporate organisations.

A well-used source for analysing social and economic issues and patterns within cities and urban areas is population census data available from national statistics bureaus or organisations. This data is based on 100% of the population (or very close to it) and is, as suggested by Batty [28], 'the gold standard for finding out what is happening in cities' (p.9). The first national census of Australia was conducted on 2 April 1911, with consequent censuses undertaken in 1921, 1933, 1947, 1954 and 1961. Since 1961, censuses have been held every 5 years, with the most recent being conducted in August 2016 (at the time of writing, the next Australian census will be conducted on the 10 August 2021).

The census aims to count the number of people in a place, usually a predefined geographic area, at a particular point in time. It does, however, also record information on dwelling characteristics, economic and demographic characteristics of the population and some, usually limited, information on daily life. Over and above data from the census, the Australian Bureau of Statistics also provides some regionally aggregated data including annual population estimates, small area labour market data, data relating to businesses (business registrations, insolvencies etc.) and health-related data.

The Australian Bureau of Statistics (ABS) website (see Fig. 1) (www.abs.gov.au) provides a one-stop shop for accessing census and other ABS data and publications. Census data is available to download using various online data tools including table builder which allows the researcher to construct purpose-designed data tables from a range of census output. Outside of Australia, statistical agencies also maintain online access portals that allow the user to access a particular county's census data remotely. Examples include the US Census Bureau (www.census.gov) and the UK Office for National Statistics (www.ons.gov.uk/).

Other government agencies also provide data that may be used in urban analysis. For example, the Australian Taxation Office releases aggregated data on a range of

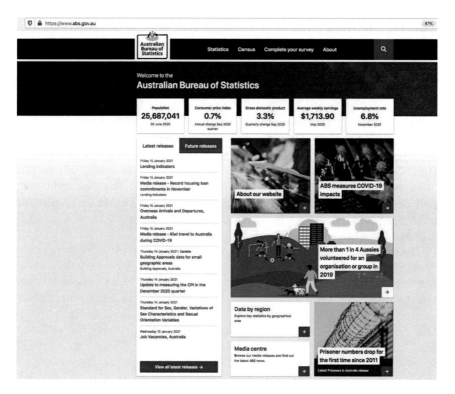

Fig. 1 Australian Bureau of Statistics home page

income tax characteristics, including total incomes, investment data, income received from government pensions and levels of student debt. The data is downloadable as a spreadsheet and is aggregated to the post-code level. Other useful spatial data available from government agencies include mortality data from the Australian Institute of Health and Welfare, data from the Australian Financial Security Authority on Personal insolvencies including why people state they have become insolvent and the causes of road deaths from the Bureau of Infrastructure, Transport and Regional Economics. All of this data is aggregated to selected spatial levels and sometimes allow comparison with census data. An excellent place to explore for spatially indexed data from government agencies is the government website (www.data.gov.au). A search for 'spatial data' returned over 17,000 results. Some of the material is simply information or contains files for use in Geographical Information Systems software, but there are also links to data dealing with issues such as the spatial location of criminal activity, health outcomes or poverty and financial stress. While most of the data are linked via external sites, the use of the data.gov.au site saves time searching a lot of different places.

Moving away from strictly government sources of secondary data, large purpose-built social surveys often have the ability to be used in urban analysis. These data sets have the distinct advantage of allowing the researcher to access a large, often national sample of respondents at zero or very small financial cost. One such Australian survey, which has comparable surveys in the USA, UK and Europe, is the Household Income and Labour Dynamics Survey (www.melbourneinstitute.com/hilda/). The HILDA Survey follows a large cohort of Australians across consecutive years, gathering responses on a variety of economic, social and labour questions. The HILDA Survey began in 2000-01 (Wave 1) and has since produced 15 consecutive waves of output, with a high rate of participant retention. Variables available from the HILDA Survey include a person's labour force status, their age, gender, health status, language command, education level, family status, their parents' employment, their level of life satisfaction as well as their social capital. Importantly, for urban analysis, the HILDA Survey includes for each respondent broad aggregate details of their place of residence. The paper on residential satisfaction by Baum, Arthurson and Rickson [19] used a combination of data from HILDA and the census to analyse respondents' level of satisfaction taking into account not only their personal characteristics but also characteristics of where they live.

The increasing sophistication of on-line data portals has led to the development of online 'clearing houses' that offer access to geo-coded data available for urban analysis research [29–31]. These portals collate spatially based data from a large array of sources and offer a public distribution mechanism for data covering a broad spectrum of research themes [30]. The Australian Urban Research Infrastructure Network (see Fig. 2) (www.aurin.org.au) run out of the University of Melbourne provides access to in excess of 1200 spatial metadata records for end users [32]. Members of registered organisations can browse and manipulate data on the portal or can download data for use elsewhere. Some of the data, such as census data, is repeated from elsewhere, but other data including information of house prices or employment vulnerability are only available from this site.

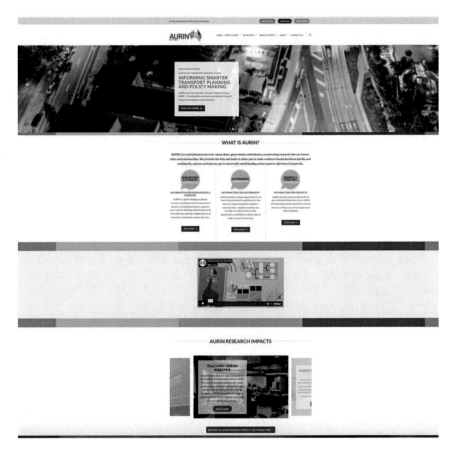

Fig. 2 Australian Urban Research Infrastructure Network

5 Big Data and Urban Analysis

One emerging source of secondary data relates to the use of big data in urban analysis [33]. The recognition and use of big data has come about due to the often seismic shift in the ways in which data is generated and processed. Schintler and Fischer [34, p. 2] assert that these shifts have been driven by:

(i) the significant improvements in storage capacity and computing power to process large data sets; (ii) the rapid increase in remote sensors generating new streams of digital data from telescopes, traffic monitors and video cameras monitoring the environment; (iii) the introduction of the internet of things, implying that even simple components and devices can communicate over the internet; (iv) the mobile revolution with the advent of mobile and smartphones enabling to receive and send information anytime and everywhere; (v) the emergence of e-commerce channels and social media platforms; and (vi) crowd-sourcing platforms for volunteered geographic information (VGI).

While definitions can vary, big data represents any voluminous amount of structured, semi-structured and unstructured data that has the potential to be mined for information used in machine learning projects, predictive modelling and other advanced analytics applications. Researchers differentiate big data from other types of data according to the 3Vs: the large volume of data in many environments, the wide variety of data types stored in big data systems and the velocity at which the data is generated, collected and processed.

Large corporations have often led the way in the use of big data in an attempt to understand customer sentiment or purchasing patterns, while it has also been used to understand and predict behaviour such as voting outcomes. [35]. Outside of the corporate world, academic researchers have also begun utilising the potential of big data [36].

In the field of urban analysis and planning research, anything that can be tagged in space or geo-coded can be used to examine patterns of location and the dynamics of these patterns [28, 37]. This includes data from social media platforms such as Facebook, Twitter or Instagram, data from online sites such as Google Street View as well as data mined from smart payment cards (i.e. Transport cards), store loyalty cards or traffic management systems). Emerging studies have included the use of social media data to study spatial inequality [38], taxi data to investigate travel patterns [39] and Airbnb listings to analyse changes in rental housing markets in New York [40].

Crowd sourcing of data has also become an interesting way of collecting data. Advances in mobile phone and other internet-linked technology have not only allowed data to be mined from social media platforms but have also enabled users to contribute to the data collection process as 'citizen scientists' by inputting data into purpose-built internet platforms. Although not at the same scale as social media usage data, the flu tracker website (see box) is an excellent example of the utility of using the public to contribute to the collection of significant amounts of data. Some examples of urban analysis using crowd sourced data include research into disaster management [41], transit ride quality [42] and daily activity patterns [43].

Small Big Data

The use of crowd sourced data to understand the city and urban areas isn't solely restricted to big data. There are several examples where crowd sourced data have been used to develop spatial presentations of particular issues. The flu tracker website (https://info.flutracking.net) is one particularly useful adaptation of crowd sourced data. The flu tracker website uses crowd sourced data from 29,000 individuals (2017) to track flu outbreaks across Australia and New Zealand and then provide maps showing flu 'hotspots'.

A more tongue-in-cheek use of crowd sourced data was used by the Australian Broadcasting Commission in 2018 who used input from listeners to develop 'The Smells of Brisbane' map which they then posted on their Facebook page.

Source ABC Brisbane [44]

Although big data is unstructured and often harder to use than other forms of data [28], it has the advantage over other secondary data sources in that it is generally collected across more regular timeframes. While population census data, which is based on 100% of the population, provides significant insight into the characteristics of an urban area (for example), it can miss significant changes in short timeframes due to its infrequent collection. As we saw above, Australian Bureau of Statistics population census is conducted every 5 years meaning that census data can sometimes miss short run changes that have occurred in the inter-census period [28]. In contrast, big data is often collected in real time providing data at significantly finer time intervals and allowing more in-depth spatial temporal investigation to be undertaken. For instance, Facebook users upload more than 10 million photos every hour and leave a comment or click the like button around 3 billion times per day, while in 2012 Twitter users were sending in excess of 400 million tweets per day [45].

6 Conclusions

The collection of secondary data covering a range of possible subject areas provides the urban researcher with an exciting body of data and information with which to undertake their studies. You can often have access to data that allows you to test and explore new research questions, without the financial and time costs of having to undertake the collection of data yourself. However, the urban researcher must also be aware that there exist potential pitfalls in using secondary data, which may need to be taken into account.

Secondary data can be accessed through any number of outlets including government agencies such as statistical organisations or through private firms and online data clearing houses. In addition, advances in computer technology have meant that the options of accessing big data are also now within the reach of the urban researcher.

Key Points

- Secondary data can be defined as data that has already been collected via primary sources and which is then made available for researchers to use for their own research.
- The use of secondary data has a number of advantages, including its general high quality, its extent and coverage and the fact that it saves the researcher time and money.
- The use of secondary data also comes with some potential pitfalls that the researcher should be away of.
- Researchers can source secondary data from an increasing number of providers, including corporations, government departments and statistical organisations. Increasingly, researchers have also stated to access big data for use in urban analysis.

Further information

For those wanting to get further information about using secondary data, there are a number of useful guides. For specific material dealing with secondary data, see

Heaton, J. (2003). Secondary data analysis. The AZ of Social Research, Sage, London, 285–288

or

T. Vartanian (2010). Secondary data analysis. Oxford University Press.

For a useful overview in the context of social science research methods, see the various chapters in:

Alan Bryman (2016) Social research methods. Oxford University Press.

References

1. Theodorson GA (1982) Urban patterns: studies in human ecology. Penn State University Press
2. Shevky E, Bell W (1955) Social area analysis; theory, illustrative application and computational procedures
3. Shevky E, Williams M (1949) The social areas of Los Angeles, analysis and typology. Publication for the John Randolph Haynes and Dora Haynes Foundation by the University of California Press
4. Abu-Lughod JL (1969) Testing the theory of social area analysis: the ecology of Cairo, Egypt. Amer Sociol Rev, 198–212
5. McElrath DC (1962) The social areas of Rome: a comparative analysis. Amer Sociol Rev, 376–391
6. Herbert D (1967) Social area analysis: a British study. Urban Stud 4(1):41–60
7. Jones FL (1968) Social area analysis: some theoretical and methodological comments illustrated with Australian data. British J Sociol, 424–444
8. Van Arsdol Jr, MD, Camilleri SF, Schmid CF (1961) An investigation of the utility of urban typology. Pacific Sociol Rev 4(1):26–32
9. Schmid CF (1950) Generalizations concerning the ecology of the American city. Am Sociol Rev 15(2):264–281
10. Berry BJL, Kasarda JD (1977) Contemporary urban ecology
11. Sweetser FL (1965) Factorial ecology: Helsinki, 1960. Demography 2(1):372–385
12. Singleton AD, Spielman SE (2014) The past, present, and future of geodemographic research in the United States and United Kingdom. Profess Geogr 66(4):558–567
13. Jones K et al (2018) Ethnic and class residential segregation: exploring their intersection–a multilevel analysis of ancestry and occupational class in Sydney. Urban Stud 55(6):1163–1184
14. Catney G (2018) The complex geographies of ethnic residential segregation: Using spatial and local measures to explore scale-dependency and spatial relationships. Trans Inst Br Geogr 43(1):137–152
15. Baker E et al (2016) Housing affordability and residential mobility as drivers of locational inequality. Appl Geogr 72:65–75
16. Cheong TS, Li J (2018) Transitional distribution dynamics of housing affordability in Australia, Canada and USA. Int J Housing Markets Anal 11(1):204–222
17. Randolph B, Tice A (2017) Relocating disadvantage in five australian cities: socio-spatial polarisation under neo-liberalism. Urban Policy Res 35(2):103–121
18. Lockwood T et al (2018) Does where you live influence your socio-economic status? Land Use Policy 72:152–160
19. Baum S, Arthurson K, Rickson K (2010) Happy people in mixed-up places: The association between the degree and type of local socioeconomic mix and expressions of neighbourhood satisfaction. Urban Stud 47(3):467–485
20. Townsley M et al (2015) Burglar target selection: a cross-national comparison. J Res Crime Delinquency 52(1):3–31
21. Metz N, Burdina M (2018) Neighbourhood income inequality and property crime. Urban Stud 55(1):133–150
22. Collins CR, Guidry S (2018) What effect does inequality have on residents' sense of safety? Exploring the mediating processes of social capital and civic engagement. J Urban Affairs, 1–18
23. Rew L et al (2000) Secondary data analysis: new perspective for adolescent research. Nurs Outlook 48(5):223–229
24. Johnston RJ (1976) Residential area characteristics: research methods for identifying urban subareas—social area analysis and factorial ecology. Social Areas Cities 1:193–235
25. Moore JC, Welniak EJ (2000) Income measurement error in surveys: a review. J Official Statist 16(4):331
26. Meyer BD, Mok WK, Sullivan JX (2009) The under-reporting of transfers in household surveys: its nature and consequences. National Bureau Econ Res

27. Davidsen MA (2016) From star wars to Jediism: the emergence of fiction-based religion. The future of the religious past, p. 14
28. Batty M (2014) Urban informatics and Big Data: a report to the ESRC Cities Expert Group
29. Pettit C, Lieske SN, Jamal M (2017) CityDash: visualising a changing city using open data. in international conference on computers in urban planning and urban management. Springer
30. Goodison CA, Thomas G, Palmer S (2016) The Florida geographic data library: lessons learned and workflows for geospatial data management. J Map Geogr Libraries 12(1):73–99
31. Crompvoets J et al (2004) Assessing the worldwide developments of national spatial data clearinghouses. Int J Geogr Inf Sci 18(7):665–689
32. Kalantari M et al (2017) Automatic spatial metadata systems–the case of Australian urban research infrastructure network. Cartogr Geograph Inform Sci 44(4):327–337
33. Jilani M, Corcoran P, Bertolotto M (2017) Geographic information systems: crowd-sourced spatial data. Wiley encyclopedia of electrical and electronics engineering
34. Schintler LA, Fischer MM (2018) Big data and regional science: opportunities, challenges, and directions for future research2018
35. Burnap P et al (2016) 140 characters to victory?: Using Twitter to predict the UK 2015 General Election. Electoral Stud 41:230–233
36. Connelly R et al (2016) The role of administrative data in the big data revolution in social science research. Soc Sci Res 59:1–12
37. Zook M, Shelton T, Poorthuis A (2017) Big data and the city
38. Shelton T, Poorthuis A, Zook M (2015) Social media and the city: Rethinking urban socio-spatial inequality using user-generated geographic information. Landscape Urban Plann 142:198–211
39. Liu Y et al (2012) Understanding intra-urban trip patterns from taxi trajectory data. J Geogr Syst 14(4):463–483
40. Wachsmuth D, Weisler A (2017) Airbnb and the rent gap: gentrification through the sharing economy. ResearchGate
41. Joseph JK et al (2018) Big data analytics and social media in disaster management, in integrating disaster science and management. Elsevier, pp 287–294
42. Lee D-H, Parsuvanathan C (2017) Assessing passenger feedback reliability in crowd-sourced measurement of transit ride quality. In: Intelligent transportation systems (ITSC), 2017 IEEE 20th international conference on. IEEE
43. Malleson N, Birkin M (2014) New insights into individual activity spaces using crowd-sourced big data
44. ABC. The Smells of Brisbane. 2018; Available from: https://www.facebook.com/abcinbrisbane/photos/a.114296864668.122116.59944789668/10156389605034669/?type=3
45. Mayer-Schönberger V, Cukier K (2013) Big data: a revolution that will transform how we live, work, and think. Houghton Mifflin Harcourt

Conducting Survey Research

Heather Shearer

Abstract This chapter is aimed at students and researchers who will use question-naire surveys in their research. It describes the basics of conducting questionnaire surveys. It begins by exploring some reasons why a researcher or another may want to conduct a survey. It then describes different types of surveys, describing how they can be differentiated by time and delivery method, and explains that most are cross-sectional (one-off) and self-administered. It then describes the necessary steps in survey research, particularly aligning the survey to the research question, and identi-fying the audience at which the survey is aimed (sample selection). The chapter then details the various aspects relevant to survey and question design (closed and open questions), including how not to write survey questions. Good question design is integral to a successful survey and this is where most surveys fall short. The different types of closed survey questions, such as dichotomous, nominal, rank order and Likert scale, are discussed, with examples of each. Some logistics around distributing surveys are given, concentrating on online survey distribution, as this is the most common method used nowadays. Finally, the chapter concludes with a brief piece on survey analysis.

1 Introduction

Questionnaire surveys are among the most used methods to collect data for urban analysis and planning research. Beginning in your student days (or even high school), anyone engaging in urban research will almost certainly be asked to create and run a questionnaire survey. Everyone will also encounter hundreds or even thousands of surveys, even if just Facebook quizzes. All surveys have the same purpose—to collect information. This information may be used for academic or market research, profiling, political polling, even extensive government exercises like the census. It may also be used for 'nefarious' purposes—for example, many social media 'quizzes'

H. Shearer (✉)
Griffith University, Gold Coast, QLD, Australia
e-mail: h.shearer@griffith.edu.au

© Springer Nature Singapore Pte Ltd. 2021
S. Baum (ed.), *Methods in Urban Analysis*, Cities Research Series,
https://doi.org/10.1007/978-981-16-1677-8_5

are deliberately created to sneakily collect personal information, sometimes even for criminal purposes.

Even though surveys are so common, many aren't perfect. Even experienced academics (and students) can create a poorly designed survey instrument that fails to answer even the most basic questions. They are too long, have poorly written questions, require too long and detailed answers, or are just plain dull. This is the reason why surveys often receive inadequate responses and low completion rates. Faced with 17 pages of open-ended questions, even the most dedicated survey fan will likely balk at completing it.

Designing a good survey requires comprehensive knowledge about data collection, research design and methodology, sampling and statistical analysis. It also requires some basic knowledge about human behaviour. Surveys are mostly quantitative (about numbers) but also collect open-ended or qualitative data (although to a lesser extent than traditional qualitative methods such as interviews or focus groups). There are many different types of survey and survey questions; and what kinds you use depend on the question/s that you, the researcher, want to answer.

Surveys generally fall into two main types [1]. First are very large and expensive data collection exercises by a government authority such as the Australian Bureau of Statistics (ABS) conducting the five-yearly census; various levels of government elections or referenda; large organisations such as News Ltd, Roy Morgan, the Essential Poll and others conducting political and consumer polling surveys to gauge voting intention or what are the most popular consumer brands; or a large institution such as a university conducting teaching experience or graduate outcomes surveys. Secondly and more commonly are the smaller research projects or informal surveys by an individual or group, such as a single student posting a social media poll asking fellow students where to hold an end of term catch-up. This chapter only discusses the latter, as large-scale survey research requires resources way beyond what a student or even a university researcher can access.

Apart from the extensive surveys that use dedicated websites, automated telephone calls, and other types of mass data collection, most surveys are self-administered (filled in by the respondent without an interviewer [2]) although a small survey is occasionally administered by a researcher in person. Nearly all surveys nowadays are run on various types of online survey software, such as SurveyMonkey [3], Google Forms, Lime Survey [4] and Qualtrics [5]. Some, such as SurveyMonkey and Google Forms, offer free access but have limited functionality. However, most dedicated survey software requires a license or subscription and is generally used only by large organisations such as universities and marketing companies because of the cost of these licenses.

This chapter aims to give some easy-to-understand tips and techniques on how to conduct survey research, specifically focused on the Australian urban research/planning framework. It will describe some common types of survey, where they are most effectively used, and how to write good survey questions. It will also explore some pros and cons of the different approaches and gives examples of good and bad survey questions. It ends by briefly describing some methods of survey analysis. It will do this by giving five different scenarios in urban planning and how a survey may be used to answer questions relating to each scenario.

2 Why Conduct Surveys?

The most basic reason why researchers conduct surveys is to answer a question or series of questions. Surveys may be used to profile or describe a sample or population; for analysis and prediction of the influence of specific variables; and developing and testing scales [6]. In urban research, surveys are mostly used for description and analysis, while developing scales is more common in psychological and demographic research. For example, the ABS's Socio-Economic Indices for Areas (SEIFA) are created to rank areas based on four scales of relative socio-economic advantage and disadvantage [7].

Urban researchers create surveys to explore aspects of the urban environment, particularly the human aspect of the urban environment. In countries such as Australia, researchers can access a wide range of data on the urban environment, such as land use, planning scheme zones or public transport infrastructure. Still, data on how the metropolitan area's population responds to the urban environment often requires a research method such as a survey. If a local council wants to change the planning scheme or revitalise some city parks, then a survey can determine what a sample of the population thinks about it.

3 Types of Survey Research

The main types of survey research can be differentiated by the delivery method (self-administered—questionnaire; or interviewer-administered—interview or focus groups) and time (cross-sectional or longitudinal). Figure 1 shows how the dimension of time can be differentiated.

3.1 Method of Delivery Pros and Cons to Different Data Collection Methods

There are two main survey delivery choices: self-administered (either online, by post or using a drop off-pick up method) or interviewer-administered (face-to-face as in interviews or focus groups, or by telephone). This chapter will mainly describe the former, but of note, surveys administered in person are more expensive and time-consuming but can garner more detailed and nuanced information. Face-to-face

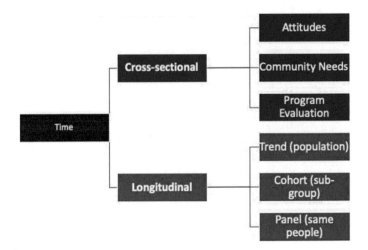

Fig. 1 The time dimension of survey research

surveys can also more finely target participants geographically or demographically. However, most questionnaire surveys are designed to be self-administered and are generally presented either online or in hard copy format.

Self-administered surveys can be sent by mail or links emailed to potential respondents or posted online. In the past, surveys were mostly sent by mail, but nowadays almost all are online. Online surveys are quick, cheap and easy to analyse, and can reach vast numbers of people and are not subject to interviewer effects (i.e. reluctance to answer certain types of questions honestly if in person). However, they can have low response rates and although better than postal surveys, poorly designed ones may be abandoned or only partially completed. Also, you have no real way of knowing how many people saw your survey, so it is almost impossible to calculate response rates. However, they are useful in getting people to fill out open-ended questions, as a respondent can type directly into the survey form (much easier than writing by hand).

They also need to be relatively short to get audience interaction and need to be designed so that the same person cannot complete it multiple times. Most survey software allows users to tweak settings, such as sending a respondent to a completion page (known as *skip logic*) if they answer a question in a specific way (i.e. are aged under 18), but as it is also an issue with postal surveys, this relies on the honesty of the respondent, so there is no way of checking if a person is who they say they are. A lesser disadvantage (mainly in developed countries) is that some populations do not have access to the internet. If wanting to target a geographically or socially disadvantaged population, it is more advantageous to use interviewer-administered surveys.

Postal surveys are useful for geographical targeting, but it is almost impossible to get a good random sample, and they are costly. It is also difficult to get names and addresses as few people nowadays use the old 'white pages' telephone directories or are even listed. Postal surveys are very time-consuming and generally require additional personnel for data input and analysis, while online software does this automatically. Mass deliveries of hard copy surveys (such as the census, which is hand-delivered to all households in Australia) are usually conducted only by large organisations or government agencies.

3.2 Time of Delivery (Cross-Sectional Survey Research)

Cross-sectional survey research is generally conducted for three purposes: to explore attitudes to something, identify community needs (and gaps) and for program evaluation. For example, a survey to understand how many people use urban parks, and for what reasons, is an example of all three; it aims to identify community attitudes to parks, what are the perceived gaps (unfulfilled community needs) in park maintenance, infrastructure, safety and the like, and program evaluation (e.g. to plan the frequency of maintenance).

3.3 Time of Delivery (Longitudinal Survey Research)

Longitudinal survey research is less common than cross-sectional research, mostly because it is expensive and time-consuming. In brief, however, trend surveys sample a cross-section of the same population (not necessarily the same sample) over time to investigate changes; cohort studies sample the same group over time (e.g. undergraduate students in a specific class every year) and panel studies sample the same people from the same population over time [8].

In urban research, researchers may want to explore urban dwellers' attitudes to something like water use or greenhouse gas emissions. In this case, they would survey a cross-section of the city's same population at regular times to see if their attitudes change. For a cohort study, a researcher investigating planning education and graduate outcomes might want to select a sub-sample, such as a specific university class, and survey this class every year. Finally, the most expensive and time-consuming are panel studies, where the same people are surveyed every year. The most well-known of these is the 'Up' television series which studied the same group of people every seven years, throughout their lives [9].

Longitudinal survey research can provide beneficial findings, but it is incredibly time-consuming and thus expensive. Moreover, there is often a high drop-out rate for panel studies, and they rely on extremely rigorous documentation and data collection (researchers themselves may drop out of a study).

4   Necessary Steps in Survey Research (Key Points/Steps in Using an Approach)

Before starting a survey, some key points to address are:

- What is your audience

    - How are you going to sample the audience?
    - Do you need Ethics approval[1]?

- What is your research question/s?
- How will you administer the survey?

    - Online, hard copy, in person?

- How many questions will you ask?
- What type of questions will you use?
- How are you going to analyse the survey?

4.1   What Is Your Audience (Sample Selection)?

What is your target population? Your research question determines this. If you want to investigate urban park users in the Gold Coast, (Australia) then people who live in the Gold Coast council area are your target population. Suppose you're going to explore the vulnerability of older women to homelessness. In that case, your target population is women above a specific age—although of course, you will likely want to narrow this down further to a country, state or even a single city.

Once you have determined your audience, you need to work out how you will sample that audience. So, assuming you do not have millions of dollars to spend, you will have to survey a smaller sample than *all* park users on the Gold Coast or *all* women in Australia over a certain age.

There are two main types of sampling methodology: probability and non-probability-based. Probability-based sampling includes:

- Random sampling
- Stratified
- Systematic
- Cluster
- Multi-staged cluster

Non-probability-based sampling includes:

- Quota

---

[1]Note, if you are a student or work for a university, all survey research requires Ethics approval.

- Convenience
- Purposive

Probability-based sampling is mostly used in quantitative research when a researcher wants a statistically robust result from which they wish to make inferences. This usually is the type of research you read about which says, '…Population prevalence of clinically significant levels of mental distress rose from 18·9% (95% CI 17·8–20·0) in 2018–19 to 27·3% (26·3–28·2) in April 2020, one month into UK lockdown' [10]. Most traditional hypothesis-based research (such as in the biomedical and environmental sciences) uses random sampling. With random selection methods, *any member of a population has an equal chance of being chosen.*

A sampling frame is a list of all the things that make up your sample; for example, it might be members of the undergraduate class 'Statistics 101', or all people living in a specific suburb or single women aged over 55 in rental accommodation. Your sampling frame also needs to exclude all individuals you do not want to survey, such as married women [11].

Sometimes a researcher may not want to choose every member of a population, so if wanting to investigate a specific sub-demographic (say, women over 55), they will stratify the population by age and gender. They may well also select participants from the other strata (for comparative purposes). Systematic sampling is a mathematical technique used to select participants, for example, if a researcher wanted to sample 500 people out of a population of 10,000. They would divide 10,000 by 500 and starting from a random number, select every 20th person. Cluster sampling is used to determine smaller clusters within a wider geographical region, such as a small cluster of Australian universities requiring students to do work experience or placements as part of their degrees. Multi-stage cluster sampling is more complex, whereby a hierarchy of groups is selected; in the above example, first, a random sample of universities might be chosen, then a sample of courses within those universities and finally a sample of students within the selected courses [2, 12].

Non-probability sampling is mostly used in qualitative data collection when a researcher wants to explore a range of different concepts. Quota sampling involves choosing pre-determined numbers of respondents according to some criteria; for example, a researcher might want to survey 100 people, comprising 50% dog walkers and 50% joggers. Once 50 joggers have been surveyed, then no more can be chosen, and similarly for dog walkers (and they may decide to exclude joggers with dogs). This can also be clustered, so if I intended to investigate five parks, I might survey 10 of each type in each park. Convenience sampling is just that; I could survey the first 20 people I encounter in each park or the first 20 who looked friendly. Purposive sampling is like quota sampling but is more specific; for example, I might want to sample 20 female dog walkers with more than one dog. Snowball sampling is one of the most commonly used methods in urban analysis; for example, I posted the links to my tiny house surveys on Tiny House social media (Facebook) pages and asked people to forward it to their friends, and also emailed it to people with the same instructions.

4.2 What Is Your Research Question?

An essential part of designing a questionnaire is in the operationalisation of your research question. This should inform the whole survey. Unfortunately, many surveys are just 'fishing expeditions' [13], asking dozens of questions in the hope of finding meaningful results. When I did my Ph.D., I looked for reasons why some households used less water than others during a major drought. The first draft of my survey asked far too many questions, ranging from behavioural theories, to childhood water use behaviour, to the presence of water infrastructure like swimming pools or water tanks, to opinions on water restrictions. This was intended to be exploratory research, but in retrospect, even the final survey had far too many questions. Fishing expeditions, the garden of forking paths, p-hacking, multiple comparisons, etc. are all similar statistical concepts [14]: I will not detail these here, as they are subjects for another chapter, but suffice it to say, you need to ensure that the survey answers the research question/s that are set BEFORE you start your survey. Restricting the survey to answering the research question/s will help keep it to a manageable length too.

4.3 Survey Design

Survey design obviously depends on what you want to find out (aka the research question or questions), who is conducting the survey, the sample being surveyed, the distribution method and the medium used for the survey. For example, designing a simple 'fun' survey as are found in their multitudes on social media (usually for 'sneaky' personal data collection or even social media phishing) [15, 16] would involve a completely different design than a detailed and lengthy market research or government survey. For our purpose, as students and researchers, we will be designing something in between, possibly taking around 15–20 minutes to complete, and aimed at answering a specific research question or questions.

So, what is the optimum length for a survey? Well, how long is a piece of string? Seriously, online surveys should take no more than 12–15 minutes to complete, and paper surveys should be 1–2 pages for a short survey and 3–4 pages for a long survey. I personally enjoy doing surveys, but even I balk at lengthy surveys, particularly those with lots of text input and open-ended questions. You should also aim for a response rate over the (relatively low) 25%, which is considered acceptable for a survey. If poorly designed, it is more likely that people will begin the survey and stop halfway through or skip entire sections [17]. These are then considered incomplete samples—missing data—and must be excluded from most analyses.

Some hints and tips for survey design include [16]:

- Keep the design uncluttered and straightforward (especially online surveys).
- The presentation should be clear; don't mix and match fonts, type sizes or add irrelevant graphics or multimedia. You don't want to distract your respondents.

- Use skip logic to target specific respondents. If you only want people over a certain age, then set the first question to ask 'How old are you?' and if they answer, for example, 'under 18', the survey directs them to the end page, and if not, they continue with Question 2.
- Don't require respondents to answer *all* questions. Do require them to answer a question such as the example above (particularly if your Ethics approval necessitates it) but if you make *every* question required, people are going to get annoyed and not finish your survey.
- Try and use different question types so that respondents don't get too bored.
- End with demographic questions (age, gender, marital status, etc.) unless required for filtering your respondents or for Ethics reasons (i.e. adults only).
- Provide a cover sheet with clear instructions on responding (and clear instructions for individual questions—you'd be surprised how many things you think are obvious are not).
- Write simply and clearly in plain English (do not assume that what *you* think is easy to understand, is indeed easy to understand). To help with this, it is useful to pilot the survey among a smaller group (but don't send them the final survey, as it will be biased).

4.4   Question Design

Good question design is integral to a successful survey. A researcher asks questions for various reasons, including to collect personal information about them or others (age, gender, income, household composition and so on); to find out about attitudes, beliefs, norms and values (what they feel about something such as political orientation, the behaviour of others); and to find out about knowledge. There are two main types of question: closed (structured) or open-ended (unstructured).

4.4.1   Closed Questions

In a closed question, respondents are given a list of choices to answer, and can only tick the provided alternatives, although it is common to include an 'other'. Closed questions produce quantitative results and are relatively easy to code and analyse. It is simple for both the interviewer and the respondent/s to answer closed questions, and they give consistent and comparable results. Response choices also make it easier for some respondents to understand the question. Closed questions are also useful to categorise respondents (i.e. 75% of respondents who said they would like to live in a tiny house were women over 55).

Closed questions have some disadvantages, especially in that responses are not exhaustive. Some respondents may find that forcing an answer is irritating because they don't agree with any of the choices. Adding an 'other' choice is one way around this but can add to the complexity of analysis. Question choices can also be biased by the researcher, who wants to explore a specific hypothesis. For example, in exploratory research, such as in the Tiny House Case Study (below) question choices were limited to the researcher's ideas why people chose to live in tiny houses, and while the question included an 'other' choice, some left this blank.

Case study—Tiny Houses

I conducted two surveys on the then-new tiny house movement in Australia. At the time of the first survey in 2015, tiny houses were not as 'big' on social media as they are today, which was likely reflected by the low numbers of respondents (56). Subsequently, in 2017, I ran the same survey, and this time, had 369 respondents. The survey was distributed via an online link on social media (Facebook tiny house group pages), thus was not random. Both surveys used the same questions, for consistency, although, in retrospect, the questions could have been improved. Nonetheless, the results of the study, when compared with other surveys on the tiny house movement, are relatively similar. Most people interested in tiny houses are older, single women and the drivers for this interest are generally economic (unaffordable housing), a desire for a sustainable lifestyle in a community setting, and self-sufficiency ('freedom').

4.4.2 Hints and Tips About Question Formulation [6, 18]

First and foremost, always keep your research questions in mind! Yes, there are so many wonderful things that we don't know about the world and would like answered, but your survey is not the place to answer every possible fascinating question you may have!

Secondly, ensure that you give clear instructions on how to answer the question (and format it in a nice, simple to read manner).

Other hints and tips are listed below:

Keep questions short (use KISS—Keep It Short & Simple): Doing this will help solve the other problems (i.e. it is easier to keep to one subject if it is short). While this is a continual source of surprise to some researchers, people are frequently confused by what *we* think is simple. I always pilot my surveys and never fail to be surprised that people don't understand a question that I thought was simple (also, they *always* notice spelling mistakes and typos)! The longer the question, the more confusing it can be.

Similarly, questions need to be in plain, simple English. Rambling and incoherent questions, such as 'Many women over 55 report high levels of housing insecurity and perhaps even domestic violence; agree or disagree?' Avoid using jargon or technical terms (don't assume that everyone has the same degree of knowledge—you may know perfectly well what a specific scientific term means, but your respondents may not—even if they are in the same discipline!) It is useful in some circumstances to give an option of 'I don't know' (e.g. 'what are the square meters of your tiny house?')

Don't ask leading, pushy or assumptive questions: Examples of leading questions are, 'What did you enjoy most of the awesome range of services in the park?' This is biased, as the question assumes the services are 'awesome'. Marketing surveys are frequently leading, pushy and assumptive. Assumptive questions such as 'What do you love about tiny houses?' assumes that you love tiny houses, when you may hate them, or not even know what they are. Also, they are annoying, and people frequently refuse to answer them.

Offering a limited range of choices for a complicated question risks pushing respondents to a (usually favoured) option. This may be something like, 'Tiny houses are a good solution for homelessness—Yes/No' or 'Employers expect planning grad-uates to know how to conduct Development Assessments?'. These are pushy and too complicated.

Don't ask random questions: Ask questions that are only related to your research question (and necessary demographic information). If I had asked respondents in the tiny house survey, 'what is your favourite colour of tiny house?', it would not be relevant to the research question (well, it *could* be, but I cannot imagine how).

Don't ask more than ONE question (double-barrelled questions): Only ask one thing at a time; for example, 'how long do you normally spend in the park and on what days of the week?' is two questions and would have more than one type of answer choice. It is common that researchers inadvertently ask such questions. I like doing surveys, and the vast majority include at least one of such questions. I always comment on this in the final text box (if available) when they ask, 'do you have anything further to add?'.

Be specific: It is essential to give enough information, such as a timeframe. For example, asking how much a respondent pays in rent is meaningless unless you specify a period (i.e. how much rent do you pay per week?) Similarly, asking a question such as, 'did you travel last year' needs to be more specific; travel where, by what means, for what purpose, and so on.

Loaded Questions

These are (thankfully) quite rare, but some rather extreme examples would be, 'When did you stop drinking to excess?' or 'When did you last exceed the speed limit'? Loaded questions are very often offensive, and even if not, are likely to be misunderstood.

Don't ask questions that breach confidentiality.

Such questions include giving a person's name or address, or other aspects related to privacy (especially around sensitive subjects like sexuality, race, or information that might put someone into danger, such as illegal drug use). Such questions *may* be asked in a survey, but it is necessary to ensure that the appropriate clearances for higher-level Ethics approvals have been granted.

4.4.3   Types of Survey Questions (Closed Questions)

There are various types of closed survey questions, which are discussed below:

Dichotomous questions: *Example: Did you vote in the last Federal election (yes/no)?*

Ensure there is an 'other' option if relevant

Nominal questions: *Example: What is your gender (male/female/other)?* Make sure the categories are mutually exclusive, and there is not more than one answer possible (i.e. age 0–4, 5–9 and not 0–5, 5–10). Ensure there is an 'other' option if relevant and that the answers are consistent with the question–if you ask, 'do you often eat vegetables'; don't then list things like 'daily, once a week,' etc.

Rank order questions: *Example: Rank the following professions according to how much you trust them, with no. 1 being the most trusted (judge, politician, doctor, teacher, journalist, car salesperson)*

Try and base the literature choices (this type of question can be *very* biased as the items on the list depend on the researcher who may choose to include more or less of what they consider 'trusted' professions). A ranking question does not let you determine the strength of preference, just the order (e.g. a respondent may detest the top 5 'hated foods' and only mildly dislike the next 5). If there are too many choices, people get confused and just list them randomly (or use techniques such as alphabetical order). Don't give more than ten choices.

Likert scale rank questions: *Example: Do you think that climate change poses a personal risk to you (Choose from 1 strongly agree, to 5 strongly disagree)?*

Ensure the scale is balanced (i.e. 'strongly disagree' should be matched with 'strongly agree'). I answered a survey a week ago, and of a 7-question Likert scale, only one was negative (the rest were varying levels of positive). You may assume that 'everyone loves puppies', but they may very well not.

Try not to use more than five categories (including a middle category of 'neutral' or 'neither agree nor disagree'). Note, however, that if a question is poorly formulated or targeted, you will get many middle category answers. I completed a survey about a trail run in which I took part, and the questions were along the lines of 'the XYZ trail run had a major impact on my life?'. I answered 3 (neutral) to nearly all of them (it was a trail run, which was hardly life-altering unless I had broken my leg or something, which I didn't)!

If you are using specific psychological theories, they tend to have very structured type of questions; arguably impossible to answer (i.e. the Theory of Planned Behaviour asks questions along the lines of 'I think it is good to do X', 'most people who are important to me think it is good that I do X', and 'I would find it easy to do X'.)

Multiple choice lists: *Example: How many of these foods did you eat last week, tick any that apply (chocolate, potato crisps, sweet biscuits, hot chips, other sweets/lollies, cake)*

Ensure that you set up the question, so people can answer more than one. You would be surprised how many surveys forget this! Such lists are also subject to bias, as the items on the list are dependent on the researcher's knowledge and could include or omit important choices.

4.4.4 Types of Survey Questions (Open-Ended Questions)

Closed questions generally comprise the bulk of most surveys, but surveys usually include at least one open-ended question. Open-ended questions are beneficial for getting qualitative and more nuanced information. They also allow respondents to answer questions in their own language and not force answers from a pre-determined list. They are useful for exploratory research and simplify conducting multimethod research. They also suit respondents who may feel more comfortable writing about sensitive details than speaking in person or on the telephone.

Open-ended questions have some disadvantages, however. You cannot automatically code (and analyse) open-ended questions as you would with closed questions. With open questions, you code them as you would any other qualitative data (mostly just using content analysis). It is, however, much more time-consuming, and you either must do it manually or use another type of software, such as NVivo. One time-saving advantage is that, unlike interviews or focus groups, open-ended questions in surveys are already transcribed (unless the survey has been a self-completion one on paper, although these have become much rarer nowadays).

Open questions (qualitative results): Example: How can the Federal Government improve response to COVID-19 crisis?

These can be part of a closed question (by adding an 'other' text box). Most software allows you to limit the text input in most software, but I tend to allow respondents to write as much as they like (this way, you can get some excellent comprehensive qualitative information).

Don't use too many open-ended questions, as a survey comprising mostly this type of question will have a high drop-out rate. I've often begun surveys that were entirely open-ended questions and gave up in the first page or two. Be very clear what you are asking in these questions (see the section on closed questions). It is easy to write confusing open questions. Use the same KISS principle.

4.5 Distributing Your Survey

Now that you have designed your survey, it is a good idea to pilot it. Piloting (or sending a draft survey to a few people) is generally very useful to identify any poorly formulated questions, survey length, comprehension and so on. I usually send it to friends, family and colleagues. Ask people to do your survey and be critical (don't worry, usually, when you send stuff to people to comment on, they love to be critical). Don't take it personally. If a friend doesn't understand a question, it is likely that any given survey respondent won't either. You don't have to take all suggestions. In fact, I would recommend that you ignore any suggestions to **add** questions (as, hopefully, you might have designed the survey only to answer your research question and not that fascinating new subject you just read on your Twitter feed).

Let's assume you will distribute your survey online, and that you are not looking for a random sample. For example, you may want to research people interested in tiny houses or residents' attitudes to urban planning. For my tiny house survey, I posted the link on social media sites dedicated to tiny houses, mostly Facebook groups and pages. (Note, you have to join a group to post, but usually can post on pages freely, depending on the administrator restrictions, if any.) I also emailed the link to people I knew were interested in tiny houses and asked them to forward it to other people. This was both a convenience and a snowball sampling method. I left the survey open for about a month but posted reminders on the sites after two weeks. This is usually necessary because social media moves very fast, and the link vanishes almost immediately.

If you were trying to find out what attitudes residents had to urban planning issues (e.g. how they would view an extensive development application in their suburb), you could post a link to the survey on social media, especially if the area has a local community group (these are usually for smaller areas rather than an entire city, but you could post it on lots of community groups). However, in some communities and certain demographics, fewer people are likely to see something posted on social media. Therefore, you could also put a link in the local government newsletter (which would require you to get permission from Council) or even hand out paper surveys at a local supermarket. Council newsletters or paper surveys would enable you to calculate response rates, as these newsletters have known distribution lists (and of course, you can count the number of surveys handed out).

5 Survey Analysis

Once you have completed your survey and collected the data you have to undertake some analysis. With most survey data, it is not simply about jumping right in and running statistical analysis. Rather, there are a number of important steps which must be carried out prior to the actual analysis.

5.1 Cleaning

The first step is to download your data in a useable form. Most survey software allows downloads in CSV format, which you can open in Excel or even SPSS (usually only with a paid subscription). Data cleaning can be the most onerous part of survey analysis (excluding qualitative transcription and coding), but it is essential aptly described by the old computer programming acronym GIGO (Garbage In, Garbage Out).

The first thing I do after downloading the data is to get familiar with it. Look over it a few times, see how the people have answered. Are there any apparent anomalies that you can see? A recent survey I created for a second-year class on urban analysis had a five-part Likert scale question (strongly agree to strongly disagree) stating '*Urban parks are a vital component of a city*?' Everybody answered either strongly agree or agree, except one person, who responded strongly disagree. This was immediately obvious, and raised the question, did the person understand the question, or did they tick the wrong box? It turned out that the respondent was my teaching assistant who had deliberately answered this way to encourage the students to pick up anomalies in the data.

If a survey respondent has many anomalous answers, it may be deliberate, but it may also be because they did not understand the questions. In this case, it may be necessary to remove that record from your data. This also applies to responses with too much missing data.

5.2 Coding

Before beginning to analyse your survey, it is essential to consider how you will code your survey. Coding can be differentiated into pre-coding (where software such as SurveyMonkey will code your questions for you) and post-coding (where you code open-ended questions after you get all the responses). Post-coding will be briefly discussed following the section on Basic Analysis.

It is essential for pre-coding to set up a master coding document, such as a spreadsheet in Excel or table in Word. When you download the data into Excel, the online program may only output coding numbers and not text. This is particularly important for Likert scale questions, where questions may be differentially coded (e.g. in one, 'Strongly Agree' might be coded five and in another, coded as 1). Unless this is noted somewhere, you may forget and incorrectly assume the opposite sentiment to what you intended. You also may wish to recode your answers; for example, by grouping ages into bins (bands). I recommend that you download the full expanded questions and code it afterwards in Excel or SPSS with a master coding worksheet.

Before you code actual answers, go through and tag all missing data with a standard code (-1 is commonly used if you are going to do quantitative analysis; alternatively, you can use something like '99'). Whatever you use, it is essential to code this appropriately in the statistical software (no data $= -1$ or whatever). You can just search and replace in most software programs or use the SPSS 'recode into the same variables' function. If some surveys have too many missing answers, it is probably better to delete these entirely (from a copy, not your original data). What proportion of missing answers should lead to the deletion of a respondent is up to you, but if they have only answered a couple of questions, it is unlikely that their answers will provide much useful information (you may still be able to use some for basic demographic statistics). It is essential to understand that some types of analysis, like regression, require that you have no missing data. It is beyond this chapter to go into the finer details of what to do with missing data, but some more information and additional references are detailed in Dong and Peng [19], in the Amsterdam Public Health Research Institute [20] and other online sources.

Note: It is important to download your data in a manner that you can interpret afterwards.

5.3 *Basic Analysis*

This is not a chapter on statistical analysis, so it will only discuss basic statistical methodology. In addition, online tools like SurveyMonkey will do a lot of simple analysis for you. Moreover, unless you are working on a very complicated hypothesis, the simplest statistical analyses (frequencies, means, etc.) are generally adequate. Some surveys are used to construct scales by using methods such as principal component analysis. However, we won't go into how to calculate scales (or factors) as it is quite complicated.

Assuming you are using SurveyMonkey and have a basic (unpaid) subscription, you can carry out simple statistical tests and charting within the program, and then download your data into Excel for further analysis. Survey results are generally downloaded as one row per respondent, with the questions as column headings (see Table 1). Download these with the questions expanded and in full (it will come up as a weird format, but you can fix this in Excel or SPSS). I usually clean my Excel data first (as it is easier to filter and find mistakes than it is in SPSS or another statistics software) and then import it into SPSS.

After cleaning and coding, the first thing I usually do is scan the data and start to recognise some of the patterns (don't worry if you are not naturally good at that sort of stuff, there are tools to help you). It's also a good idea to make a copy of the data and do the analysis on this copy (so you can keep the original 'pristine').

The following will briefly describe some basic principles analysing survey data:

First, it is important to understand measurement levels; is your data **nominal** (discrete categories such as fruit, gender, dogs), **ordinal** (ranks such as Likert scales, ratings), **interval** (numbers with no zero, such as temperature) or a **ratio** (true numbers)? This information and how many variables you are analysing determine the

Table 1 Simple univariate and descriptive statistical analysis (adapted from Raghunath [21])

Level of measurement	Examples	Statistics (descriptive)	Type of chart
Nominal	Fruit, animals	**Dispersion** (frequencies, percentages) **Central tendency** (mode)	Bar, pie
Ordinal	Likert scales, GPA	**Dispersion** (frequencies, percentiles) **Central tendency** (mode, median)	Bar, pie, stem and leaf
Interval	Temperature, some Likert	**Dispersion** (frequencies, SD) **Central tendency** (mode, median, mean)	Bar, pie, stem and leaf, boxplot, histogram
Scale	Weight, money	**Dispersion** (Standard Deviation (SD)) **Central tendency** (mode, median, mean)	Histogram, boxplot, line

type of statistical test you can do. Table 1 shows some simple statistics for descriptive, univariate (one variable) analysis, and for more information, Goldstein [22] gives some useful tips.

If you are just doing descriptive research, you only need to calculate frequencies, standard deviation and mean mode or median if analysing more than one variable (bivariate or multivariate analysis of variables and groups (multivariate), some of the more commonly used inferential/multivariate statistical measures are contingency tables (or crosstabs), chi-square, correlation, regression and analysis of variance (ANOVA).

Note: These statistical measures only measure relationships and not causality—there are many examples of statistically significant relationships between completely unconnected variables—what some term 'spurious correlations' [23].

- Contingency tables (crosstabs): This is a very flexible method used for any pair of variables (e.g. the gender of the respondent (row) and income (column)). Using this method, it is easy to see patterns in the data.
- Chi-square: This method is used for nominal or ordinal variables. This test is applied to contingency tables to identify a relationship (or not) between variables. It compares the expected value of a cell (based on chance alone) and calculates a statistic based on the cell's actual value. The expected value is based on the data and not any literature!
- Correlations: The correlation coefficient (usually Spearman's or Pearson's depending on your data—don't worry, the program does it for you) calculates the statistical likelihood of a relationship between any number of variables. Of important note, however, the larger the sample, the more likely you will find significant correlations.
- ANOVA: ANOVA is frequently used in the physical sciences and in survey analysis, to compare the means of different groups. There are various types of ANOVA; but for a simple one-way ANOVA, you need at least three groups of a categorical dependent variable and an independent continuous variable. For example, for case study 4 (older women and homelessness) you might compare the women grouped based on a previously categorised type of homelessness (i.e. living with children, housesitting, living in temporary accommodation) against an independent variable such as income or age. ANOVA compares the means between groups and within groups and estimates significance based on that calculation.
- Simple linear (or ordinary least squares) regression: regression analysis is the fundamental basis of most statistical analysis. Essentially, 'Regression is a statistical method that allows us to look at the relationship between two variables while holding other factors equal' [24]. As can be easily deduced from the names, bivariate regression has two variables (one dependent and one explanatory), and multivariate regression has many variables (one dependent and many explanatory). This type of regression analysis requires that the dependent variable is continuous (scale or ratio data). If wishing to analyse a categorial (binary, nominal or ordinal dependent variable), it is necessary to use either binary, nominal or ordinal logistic regression.

It is important to note that it is not advisable—from a statistical perspective—to calculate means on Likert series coding [25]. Likert scales are ordinal data which are 'artificially coded' into numerical data. The problem with Likert scales is that there is no real number attached to 'Agree' or 'Strongly Disagree'. You cannot average 'Agrees'. What you can do is count them and work out frequencies based on another variable (i.e. 74% of tiny house dwellers' strongly agreed 'that affordability was the primary motivator for them building a tiny house'). In this way, you have a real number of tiny house dwellers, and you can see the frequency with which each Likert scale option was chosen.

Moreover, it is also important to understand the type of statistics that you are going to use. In the past, students had to manually calculate many types of statistics (even multivariate regression), but nowadays there is software that will do everything for you. You must bear in mind the old computer coding phrase GIGO, which means Garbage In, Garbage Out. You can put any old data into something like SPSS, and it will return a result, often 'significant'. Unless you have carefully cleaned and fixed up your data and used the correct statistical test, this result will be meaningless.

5.4 Analysing Qualitative (Open-Ended) Data

Finally, we will briefly discuss analysing qualitative data [26]. There are many methods used to analyse qualitative data, including grounded theory, content, discourse, narrative and framework analysis; and most are generally used with large amounts of data. Given that surveys typically only have a small proportion of qualitative data (as mentioned above, we try and limit open-ended questions), and that qualitative data analysis is sometimes complicated, I will only discuss the most basic qualitative coding method.

Qualitative coding is part of content and other methods of analysis, and it serves to reduce a large amount of text to a smaller number of themes. Thematic coding is generally done with a combination of manual analysis (multiple readings of the text) and automatic analysis using software such as NVivo. It is also common that more than one researcher will code the text and iteratively develop the resulting themes. It is qualitative, and thus subjective. Different researchers, therefore, may derive entirely different themes from the same data.

Very briefly, I will cut and paste text from the open-ended questions into Word or Excel and then upload this to NVivo (this can accept almost any data, including multimedia and I use the program for many purposes, including as a reference library, as it has excellent search functionality). It is essential to read the text a few times to get an idea of the respondents' 'mood'.

Then, you can use NVivo to automatically find words that are used most commonly in the text. Unfortunately, this only finds single words and not phrases, but generally speaking, the two or so words of a phrase, say 'tiny houses', will occur with the same frequency. You can manually delete certain words (such as the or/and, set word limits and so on). In doing so, you can also copy the paragraph or sentence in which common words occur and create themes automatically. In reading the text, you can highlight and add specific phrases or quotes into an existing theme (or Node in NVivo). The software also does much other analysis (once you have derived your themes) including creating word clouds and tree diagrams of related terms.

Probably one of the most useful parts of analysing qualitative data in surveys is to get a much deeper and more nuanced idea of what people think about the issue, be it tiny houses or becoming homeless as an older person. It humanises your respondents and is also great for getting quotes for your papers!

6 Conclusions

In conclusion, this chapter gave an overview of how to conduct survey research. It began with some of the reasons why urban researchers conduct surveys, particularly for value-add to purely quantitative data, such as maps of planning scheme zones, or existing land-use. Surveys allow researchers to explore the human dimension of urban research.

We then discussed the different types of survey research and how they can be differentiated by method and delivery time. Surveys can either be self-administered or interviewer-administered. Increasingly, surveys are being conducted entirely online, which is quick and cheap and does not require transcribing or separate data input. Self-administered surveys also include hard copy questionnaires, but there are less common nowadays, except for extensive and expensive surveys, such as the Census. We also discussed the difference between cross-sectional and longitudinal studies, noting that the latter can be extremely valuable data collection exercises, but are time-consuming and expensive to run.

The chapter then explained the necessary steps in survey research including sample selection (choosing your audience and differentiating between random and non-random selection), ensuring your survey answers your research question/s, and survey and question design. Some useful hints and tips about survey design were given, including to keep it simple, reduce clutter, ensure it is not too long and test the survey by piloting it amongst your friends, family or colleagues before publicly releasing it.

Designing good questions detailed the different types of survey questions that you can ask, and how they can be differentiated into closed (limited choice of responses, mostly quantitative) and open questions (open-ended answers, usually qualitative). The section also detailed some of the different types of questions within the categories, including nominal, rank order and Likert scale questions. It gave some useful tips on what questions *not* to ask, such as double-barrelled, leading or loaded questions. It then described some ways of distributing surveys, mostly focused on online surveys.

The chapter concludes with some basic survey analysis, including the all-important steps of cleaning and coding the data prior to analysis. It listed some basic statistical analyses that can be done on survey data, and what types of questions are best suited to what types of analysis. Finally, it discussed the qualitative analysis of open-ended survey questions.

Key Points:

- Questionnaire surveys are a rapid and useful method of collecting data for urban analysis and planning research.
- Questionnaire surveys are commonly used throughout academia, government and private industry.
- Urban researchers create questionnaire surveys to explore aspects of the urban environment, particularly the human aspect of the urban environment.
- While questionnaire surveys are ubiquitous, it is important that they are designed correctly to garner a good response, and address the research question/s.
- A good questionnaire survey includes targeting the appropriate population, designing effective questions and analysing the data with appropriate statistical methods.
- While mostly quantitative, questionnaire surveys can also be used to rapidly collect good qualitative data.

Further information

For those wanting to get further information about conducting surveys, there are a number of useful guides.

For an in-depth guide try:

- Floyd Fowler (2013) *Survey Research Methods*, Sage publications,
- Christof Wolf, Dominique Joye, Tom E. C. Smith and Yang Chih Fu The SAGE handbook of survey methodology, Sage publications

For a useful overview in the context of social science research methods, see the various chapters in:

- Alan Bryman, (2016) *Social research methods*. Oxford university press.
- Earl Babbie, (2020) *The practice of social research*. Cengage learning.

References

1. Andres L (2012) Designing and doing survey research. Sage
2. Bryman A (2016) Social research methods. Oxford university Press
3. Survey Monkey. Survey Monkey, https://www.surveymonkey.com/ (2020)
4. Lime Survey. Lime Survey, https://www.limesurvey.org/ (2020)
5. Qualtrics. Customer Experience Surveys, https://www.qualtrics.com/au/customer-experience/surveys/ (2020)
6. Rowley J (2014) Designing and using research questionnaires. Management Research Review

7. ABS. 2033.0.55.001—Census of Population and Housing: Socio-Economic Indexes for Areas (SEIFA), Australia, 2016, https://www.abs.gov.au/ausstats/abs@.nsf/mf/2033.0.55.001 (2018)
8. Caruana EJ, Roman M, Hernández-Sánchez J, Solli P (2015) Longitudinal studies. J Thoracic Disease 7(11):E537–E540. https://doi.org/10.3978/j.issn.2072-1439.2015.10.63
9. Wikipedia (2020) Up (Film series). Internet website https://en.wikipedia.org/wiki/Up_(film_series)
10. Pierce M, Hope H, Ford T, Hatch S, Hotopf M, John A,…Abel KM (2020) Mental health before and during the COVID-19 pandemic: a longitudinal probability sample survey of the UK population. Lancet Psych 7(10):883–892
11. Statistics How To (2014) Qualities of a Good Sampling Frame. https://www.statisticshowto.com/sampling-frame/
12. Kalton G (1983) Quantitative applications in the Social Sciences: Introduction to survey sampling. SAGE Publications, Inc., Thousand Oaks, CA. https://doi.org/10.4135/9781412984683
13. Lewis-Beck M, Bryman A, Futing L (2004) Fishing expedition
14. Leek J, A menagerie of messed up data analyses and how to avoid them, https://simplystatistics.org/2016/02/01/a-menagerie-of-messed-up-data-analyses-and-how-to-avoid-them/
15. Crozier R (2018) Facebook reveals 63 A/NZ users took quiz behind Cambridge Analytica scandal, https://www.itnews.com.au/news/facebook-reveals-63-a-nz-users-took-quiz-behind-cambridge-analytica-scandal-488715 (2018)
16. Giles M (2017) Social Media Phishing: a Primer, https://inspiredelearning.com/blog/social-phishing/
17. Hoerger M (2010) Participant dropout as a function of survey length in internet-mediated university studies: implications for study design and voluntary participation in psy- chological research. Cyberpsychol Behav Soc Netw 13:697–700
18. Barrett J (2020) The 7 Deadly Survey Questions, https://www.getfeedback.com/re-sources/online-surveys/7-deadly-survey-questions/
19. Dong Y, Peng C-YJ (2013) Principled missing data methods for researchers. Springer-Plus 2:222
20. Amsterdam Public Health Research Institute. Handling Missing Data, http://www.emgo.nl/kc/handling-missing-data/ (2015)
21. Raghunath D (2019) Data Levels of Measurement, https://medium.com/@rndayala/data-levels-of-measurement-4af33d9ab51a
22. Goldstein, E. Choosing a Statistical Test, https://www.youtube.com/watch?v=UaptUhOushw (2016)
23. Vigen T (2020) Spurious Correlations, https://www.tylervigen.com/spurious-correlations
24. Jones B (2019) A short intro to linear regression analysis using survey data, https://medium.com/pew-research-center-decoded/a-short-intro-to-linear-regression-analysis-using-survey-data
25. Statistics How To. Likert Scale Definition and Articles, https://www.statis-ticshowto.com/likert-scale-definition-and-examples/ (ND)
26. Erlingsson C, Brysiewicz P (2017) A hands-on guide to doing content analysis. African J Emergency Med 7:93–99

Using Focus Groups in Applied Urban Research

Laurel Johnson

Abstract This chapter explores the many ways that focus groups are employed in applied urban research. Applied urban research underpins and directly influences urban practices to resolve urban issues. Urban practices can be policies, plans and strategies related to diverse fields such as transport, placemaking, area renewal, new development and other endeavours. Applied urban research is undertaken at different spatial scales in a range of settings: site, neighbourhood, district, city, city region. The surprise in this chapter is the adaptability of the focus group method to suit these various urban research contexts. By investigating the focus group method, I hope to persuade the reader to the value of the method and the many purposeful ways to deploy the technique in applied urban research.

1 Introduction

In the social scientific tradition, the focus group is a qualitative research method used to collect data from a group discussion [1]. Importantly, focus groups are a data collection method and the interaction between members is a key feature of the method. The focus group process must involve an exchange of ideas between members. The interaction between focus group members as they deliberate an issue allows an exchange of ideas that identify shared and different perspectives on the issue [2].

Urban issues are multifaceted and the ability to 'go deep' into an urban issue improves the researcher's understanding and hence the quality and complexity of analysis. Focus groups enable a depth of understanding as the back and forth exchange of ideas in the group encourages exploration of areas of agreement and disagreement that enrich the resultant focus group results [3]. In policy-related urban research, the focus group method supports a range of activities such as urban planning, place-

L. Johnson (✉)
University of Queensland, St Lucia, QLD, Australia
e-mail: l.johnson3@uq.edu.au

© Springer Nature Singapore Pte Ltd. 2021
S. Baum (ed.), *Methods in Urban Analysis*, Cities Research Series,
https://doi.org/10.1007/978-981-16-1677-8_6

making, social and economic impact assessment, social infrastructure strategies, parks and open space design, action planning, heritage protection, transport planning and others.

2 Roles and Types of Focus Groups

There are multiple roles for focus groups in urban research that reflect the complexity of urban research, particularly applied urban research for policy development. I have used the focus group format in many policy-related research undertakings for profiling, values and issues identification, vision and goal setting, strategy development, scenario testing, impact assessment, prioritisation, verification and more. The value of the focus group in these contexts is as a mechanism for groups to deliberate on issues and identify themes, possibilities, priorities and alternative actions.

A focus group takes many forms. The groups can be drawn together from various aspects of urban life. The membership of the group depends on the purpose of the urban research. I present a typology of focus groups in Table 1 to assist the researcher to select the right group or combination of groups for their research task.

While the spatial focus group primarily has a geographic dimension, in urban research, the geographic context of the other focus group types is likely to be relevant to the research. In urban research, geographic location is generally of interest to the researcher. The geographic or spatial dimension in urban research sets it apart from many other social science research activities.

2.1 Focus Groups in Policy-Focused Urban Research

The purpose of this section is to present a generic urban policy research process (in this case, for urban plan making) and characterise the roles of focus groups at each

Table 1 Focus group typology

Type	Characteristics	Examples
Spatial	Group members are resident or have a primary attachment to a bounded geographic area	Neighbourhood or local resident group, Traditional owners, local business operators
Cohort	Group participants are members of a population cohort	Young people, older people, homeless people, people living with a disability
Interest	Group members share an interest or an endeavour	A school parents and friends group, environmental group, industry group, faith group
Service user	Group members use the same service or facility	Public transport users, community facility users, open space users

stage of the process. This can stimulate ideas for focus groups in your urban research activities and projects.

Table 2 maps the stages of urban policy research for urban plan making and identifies roles for focus groups at each stage. Sometimes a focus group can scope and steer the research direction. While that is not strictly data collection for the purposes of the research, the insights generated in these focus groups add value by guiding and directing the data collection and analysis effort. This is both efficient and valuable for research credibility as it triangulates the research. The focus group is another data collection method alongside others and also a vehicle for research participants to shape the research.

3   Forming Focus Groups in Urban Research

Focus groups are an established and valuable method with a long history in social scientific and market research. Focus groups were introduced into social scientific research in the 1940s as an extension of the individual interview [4]. The strength of the focus group is in the interaction of its members for the active exchange of ideas among participants [1]. During the focus group exchanges, new understandings emerge including shared perspectives (consensus) and diverse views. There are no 'right' or 'wrong' answers, but rather exploratory discussions that are sometimes highly structured and at other times, free flowing and organic. As both of these approaches have benefits, a skilled focus group facilitator can use both structured and unstructured approaches in focus group sessions.

3.1   Recruiting Focus Group Members

In social science research, there is debate about whether to use existing groups or to create time-limited 'one-off' groups to serve the research purpose [1]. Social scientists tend to favour the formation of 'one-off' focus groups for the research as they argue that ongoing (existing) groups can be unfocused for the purposes of research [1]. In urban life, many interest groups form due to a bounded geographic area such as a neighbourhood or a local park, traditional country or the site of development. These groupings are a natural collective and are convenient for focus group research in discrete geographic areas.

In urban research, there are advantages to using existing groups for focus group research. This is because many communities form groups to represent and pursue specific interests and issues that are of interest to their members. For example, where the research topic relates to a school community, a school's parents and friends organisation can be an effective focus group. This is because the group members have deep knowledge of their school community and a commitment to that community. Consequently, they are likely to be open to engaging in research that relates to the

Table 2 Focus groups in urban policy research

Urban policy research stage	Purpose of the research stage	Focus group roles
Profiling a place and its population	• Define the scope of the research • Stakeholder analysis • Identify the individuals and groups with a likely interest in the research	• Build relationship between the researcher and the community • Introduce the project to the community • Identify and confirm research participants • Verify place and population profiling • Guide the research and scope issues in the research
Issue identification and prioritisation	• Contextualise and prioritise issues • Identify data needs • Guide data collection by identifying the variables and general causal relationships to be investigated in the research • Identify individuals and groups to engage in the research	• Verify the findings of past research and the efficacy of current policy and practices • Identify and prioritise issues and potential data sources • Direct the engagement and recruitment strategy for the research • Refine research design and research methods • Through prioritisation, the issues are contextualised to a spatial area and/or a specific population cohort such as young people or an interest group such as local business operators or service user groups such as bus or park users
Values and visioning	• Understand the stakeholder values and vision • Know the strategic setting for the research • Guide data collection by identifying the variables and causal relationships to be investigated in the research	• Identify values • Drafting and/or verify a draft vision • Seek consensus, identify concerns and collect feedback on the draft vision
Goal/objective setting	• Distil and prioritise the key goals and objectives of the urban policy/plan/project from the vision • Refine data collection with more specific variables and causal relationships to be investigated in the research	• Identify data sources • Verify data sources • Prioritise goals and objectives • Verify goals and objectives

(continued)

Table 2 (continued)

Urban policy research stage	Purpose of the research stage	Focus group roles
Alternative strategies	• Identify the alternative ways that the goals/objectives can be achieved	• Identify the criteria for assessing alternative strategies • Verify the criteria for assessing alternative strategies • Identify the implications of alternative strategies for different stakeholder groups
Preferred option or strategy	• Choose the preferred strategy that best satisfies the goals and objectives and aligns with values and vision	• Assess and prioritise alternative strategies • Verify the preferred strategy
Implementation	• Prepare action plans and seek commitment from key agents	• Audit past actions • Identify possible future actions • Seek commitment to agreed actions
Monitoring	• Ensure the policy is achieving its goal/objectives • Undertake formative evaluation of the policy	• Provide feedback on the effectiveness of the policy through formative evaluation • Verify evaluation criteria and monitoring framework • Identify monitoring data sources, data collection and distribution
Reviewing	• Conduct formative and summative evaluation of the policy	• Undertake thematic evaluation of the policy • Evaluate the effectiveness of the policy for different stakeholder groups

school. The parents and friends group can be enlisted for the research and a focus group method can be used to gather information for the research. The researcher can assume that the members know each other and that some group protocols have been established. Skilled facilitation will keep the group focused on the research topic and ensure that quiet group members are heard. In this context, it is expedient and efficient to use an existing group as a focus group for data collection. This is also a respectful way to engage community members in research, as the researcher situates themselves within the community's context.

In my research, I have engaged homework club members in transport research in a transport disadvantaged community. Young people in the group had a keen interest in transport including public transport and the walkability of their neighbourhood. The focus group process was facilitated and group members were invited to use coloured pens to draw their transport and mobility experiences (challenges and solutions) in the local area on base maps. The focus group process took just over an hour. The fcous group process caused minimal disruption to the three-hour homework club, while the research benefited from the level of trust that already existed between group members, which, in turn, enabled the open exchange of ideas and information.

A contentious issue in the focus group method is paid participation. There are ethical concerns that recruiting focus group participation through paid incentive will bias the participants, their responses and the results. Though, there is a counter view that local knowledge is expert knowledge and it should be valued with financial reward. In my experience, paid participation in focus groups in Australian urban research is limited to Traditional Owners where traditional knowledge is being shared and homeless people or other socially disadvantaged participants where paid participation is respectful. Travel reimbursement and the provision of catering are generally adequate compensation for focus group participation.

3.2   Size and Facilitation

Focus groups ideally comprise between six to eight individuals [5]. This number allows for a balance of individual input and group deliberation. Where existing groups have more than ten members and a focus group method is used, then two focus groups can be conducted. The focus group should be facilitated by a skilled group moderator. This role is generally undertaken by the urban researcher or someone with small group facilitation skills who is not a member of the focus group. The success of the focus group method centres on the skills of the facilitator. If you plan to conduct focus groups in urban research, you must undertake skills development in small group facilitation before using the method. Focus groups can be unpredictable and their effective management is essential to ensure that the data and information elicited from the group adds value to your research.

3.3   Timing and Location of Focus Groups

Depending on its purpose and context, a focus group process can be short or extend up to two hours or longer. The timing of the group (when it takes place) is best negotiated with the focus group members and the timing should be communicated to the group well before its commencement. Two weeks' notice of the focus group time, date and location is reasonable. Many existing groups in urban life have regular meetings. Consistent with being a respectful researcher, this can mean that you undertake the focus group in the group's existing processes (scheduled meetings). Your focus group could be an agenda item at a regular meeting of the group. In that case, make sure that there is enough time on the agenda for the focus group. The location and timing of the focus group/s is negotiated with participants. In some places, such as outer suburban or new growth areas, there may be limited options to locate a focus group in the area. In that case, think about using facilities at the local school or asking if focus group members can travel to a nearby venue or attend the focus group remotely/online. Facilitation of the group includes an agenda with time for introductions and a closing procedure for the group.

4   Focus Group Procedures

Whether the group is pre-existing and co-opted to the research or formed for the purposes of the research, the attention to individual and group comfort is important. Venues should be familiar places for focus group members. Venues should be visited beforehand by the researcher (where possible) to check the space for adequate and comfortable seating, table layout, risks of external noise and distractions and time of day. For example, is there adequate lighting for an evening group? Where a visit is not possible, find photographs of the venue and check the details.

Catering is also important. Water (at least), hot drinks and snacks should be available to focus group members. Sometimes, pre-briefing materials and pre-briefing contact through email or phone contact is helpful to ensure that the focus group members understand the research topic and their role in the research.

The best way to support the focus group is to be prepared. Being prepared means:

- inspect (where possible) and book a suitable venue with adequate and appropriate size, access, seating, facilities, lighting, fresh water, tea and coffee facilities, car parking and noise attenuation;
- give focus group members plenty of notice of the meeting: two weeks' notice with a reminder one week from the meeting is ideal;
- brief focus group members on the purpose and length of the focus group before and provide an agenda beforehand;
- ensure ethical clearance is achieved and ethics documents are available to focus group members and member consent forms (including consent to record and photograph as relevant) are signed before commencing.

Table 3 is an example of a generic running sheet that is helpful to prepare when conducting focus groups. The running sheet aligns with and extends the focus group agenda though it is usually not distributed to the focus group members. The running sheet assists the urban researcher and/or facilitation team to ensure consistency across focus groups, where more than one group is conducted within the research.

4.1   Recording the Focus Group

With the explicit permission of each member, focus groups can be recorded using voice recording or video devices. If this is not possible, the focus group can be recorded through note taking. In the case of focus groups with note taking, it is challenging for the facilitator to also be the note taker. It is important to allow the facilitator to manage and moderate the group to maximise participation and note

Table 3 Sample running sheet for a one hour focus group

Timing	Agenda item	Purpose	Process	Resources/Roles
10.00–10.05	Welcome and introductions	Welcome and introduce members and outline the research purpose and focus group process	Facilitator acknowledges country, introduce themselves and any support personnel and invites participants to introduce themselves	Focus group agendas are available for all to see throughout the focus group meeting All consent forms are signed (including consent to photo graph/record) Focus group member name tags
10.05–10.30	Question 1: identify the issues	Collect answers to question or issue	Invite each group member to share their response to question 1 This is generally an exploratory question to elicit issues	Small group facilitator, note taker, time keeper, recorder
10.30–10.55	Question 2: explore, prioritise, solve the issues	Collect answers to question or issue	Invite each group member to share their response to question 2 This is generally a supplementary question to prioritise or further explore the responses to question 1 or identify solutions to the issues raised in question 1 responses	Small group facilitator, note taker, time keeper, recorder
10.55–11.00	Next steps and close	Present the next steps in the research and thank participants	Explain how the focus group results influence the research and close the group process	Have contact details available for focus group members who want further information about the research or to stay in touch with the researcher

taking can be a distraction to the facilitation task. In this case, the group facilitator can be supported by a note taker who is a member of the research team and not a member of the focus group.

4.2   Online Focus Groups

Online attendance is possible for focus groups with some focus groups occurring entirely online. During the COVID pandemic, I conducted eight place-based online focus groups to identify issues and strategies to improve the economic development prospects in an urban Local Government area. Each group was conducted using the same process over a two-hour period with the same research team supporting each group. That research team was made up of two facilitators (one facilitating the online chat and the other facilitating the online conversation), a note taker to capture key issues and ideas and a timekeeper to keep the focus group process on task.

4.3   Use of Technology in Focus Groups

Whether online or face to face, the use of technology during the focus group has become a popular way to identify and prioritise issues. Voting apps allow participants to vote or input issues in real time and the collective results can be reflected back to the group for discussion during the focus group. This is particularly valuable for younger focus group members (such as school students) who can use their familiar devices to input to the focus group research topic in real time. I have used technology in focus groups for group members to present their issues in a geographic area, using simple geographic information system platforms that enable the 'mapping' of information and the generation of collective 'maps' during the focus group.

4.4   Recruiting Focus Group Members

As discussed earlier, sometimes, focus groups are formed to meet the research need and at other times, existing groups in the community are co-opted to focus their attention on the research topic. That is, there are two distinct ways to recruit focus groups in urban research. Both ways of forming focus groups are accepted in urban research and there are advantages and disadvantages associated with both approaches.

The way that the focus group forms will depend on the research context. Some existing groups such as Traditional Owners have a strong attachment to a place and ongoing groups like this should be engaged in research in ways that make them comfortable and that are respectful of group protocols. The focus group method is a way to engage existing groups in research though the research interest is not the reason for the group's existence and the focus group activity might be one agenda item among many.

4.5 *The Focus Group Within a Process*

The focus group can be valuable as a process within a meta-process. For example, I have used focus groups within larger gatherings such as conferences and even public meetings. In this context, each focus group requires skilled and consistent facilitation and a structured process is necessary to instil confidence in the comparative analysis of the multiple focus group results. In this scenario, I have used table-based focus groups with either pre-selected members or random seating arrangements. This is acceptable where the research topic is the focus of the event or public meeting as it is assumed that participants have self-selected to attend as they have an interest in the topic.

5 The Focus Group in Mixed-Methods Urban Research

Many urban research projects are mixed-method and focus group results are analysed in association with other data and sometimes to clarify, verify or prioritise the results of these other data. Focus groups can complement and support other data collection methods in urban analysis. For example, I have successfully used focus groups to guide quantitative data collection in mixed-method transport research. In one research project, a focus group of transport users identified the range of issues that constrain their travel. In turn, those issues informed the selection of variables in the quantitative analysis such as cost and frequency of public transport to desired destinations and others. The focus group results guided investigation of the cost and frequency of transport services to the destinations of interest to the local travellers.

The focus group can input to quantitative analysis and clarify and verify quantitative outputs to go deeper and reveal the reasons for the patterns in quantitative data. In a transport project, focus group members identified that there is no smart card purchase or 'top up' facility in their local area. That information underpinned the high use of paper tickets for public transport travel in that area that was identified in analysis of quantitative travel data. The layering of data, in this case, triangulates the research to enhance understanding of this issue.

6 Complementary Urban Research Methods

To maximise the useability of the focus group results, it is important that the researcher has undertaken some research before commencing the focus group processes. A literature review, including policy and media content analysis assists the researcher to understand the topic, acknowledge any past research, identify the focus group type and potential members and frame the questions for the group. This preliminary research of the topic may include key stakeholder conversations to ensure that the scope and membership of focus groups is appropriately targeted. These initial investigations prepare the researcher with an understanding of community sensitivities or conflicts that exist around the research topic.

Focus group results can identify areas within the research topic that need further exploration [1]. Research methods such as quantitative data analysis, in-depth interviews, field work and observation can then be engaged to complement and extend the focus group results. For example, individual in-depth interviews either with focus group members or others outside of the group are a useful method to follow-up specific issues that emerge from the focus group interactions.

Another complementary method is field work. For example, returning to the transport disadvantage example, the homework club focus group identified an informal path to the nearest railway station. The young people directed me (the researcher) to the hidden path. The path was unsealed, isolated and through bushland. The path was not known to the agency responsible for accessibility in that local area. The field work led to documentation of the path through photography and film. This extended qualitative research was the direct result of information that emerged in a focus group. The combination of the results from the research methods was powerful and resulted in a new, safer path being built in the local area. This is the rewarding result of employing a focus group method to guide and direct mixed-methods urban research. Table 4 summarises some of the ways that focus groups can interface with other research methods in mixed-methods urban research projects.

7 Focus Group Data: Analysis and Presentation

As the focus group is purposeful, the results are not generalisable [1]. Nevertheless, focus group data can be analysed and the results can influence the research. Table 5 is an example of the analysis of focus group data presented in a single table format. The advantage of this format is that the reader can see the theme, the frequency of focus group comments related to the theme and the de-identified quotes (selected) from the focus group participants. In this transport research, there were 5 cohort focus groups conducted and the results are combined in this single table to better understand the prioritisation of community concerns.

Table 4 Focus groups and mixed-method urban research

Research methods	Link to the Focus group
Review literature such as past research, policy and media content analysis	Equips the researcher with insight into the topic and/or the location of the research prior to the focus group. This assists to: • Scope the focus group questions • Identify focus group participants • Identify community sensitivities about the research topic that may emerge during the research, including in focus group deliberations
Quantitative data analysis	The focus group can: • Direct the quantitative data collection effort by informing the variables and relationships between the variables that matter • Verify and explain the patterns in quantitative data such as travel behaviour, socio-economic profiles and others • Expose the impacts and implications of general phenomenon (such as socio-economic disadvantage) on daily life • Reveal the ways that communities meet the challenges that are implied in the quantitative patterns such as high unemployment, high levels of socio-economic disadvantage and others
Individual interview	The focus group can: • Identify issues for exploration through in-depth, individual interviews • Identify individuals who hold particular knowledge or influence that is of interest to the research
Field work/observations	The focus group can: • Guide the researcher to locations and sites in an urban area that are significant to the research

Limitations of the Focus Group in Urban Research

The key limitations of the focus group method in urban research relate to ethics and confidentiality and the comparative resource requirements for focus groups.

Ethics, Confidentiality and Resources

Ethical human research procedures will (where possible) guarantee confidentiality and protect the research participants by de-identifying them as the source of data. This guarantee is not always possible in a focus group, particularly where existing interest groups are used or where the group is formed due to common interests or characteristics in a smaller urban setting such as a suburb. It is most likely that focus group members will know each other so, exchanges of information within the

Table 5 Example of presentation and thematic analysis of focus group results [6]

Transport theme	Frequency of mention (n = 47)	Selected quotes and observations
Access to employment and education	19% n = 9	– "I'm looking for a job at the moment. I'm going to get the job and worry about the transport later" – "If students miss the 8.30 am Westside bus to Forest Lake High, the next bus is not until 9.30 am and they get detention for being late to school" – "The Westside bus is crowded with Forest Lake high school students Wednesday afternoon as it is the one day we all leave school at the same time at 2.30 pm. We sit on the floor of the bus and on each other's laps" – "No weekly after dark, Saturday afternoon or Sunday services mean we're bored and stuck at home"
Adequacy of transport infrastructure	17% n = 8	– "Progress Road overpass to Wacol Station is narrow and dangerous for pedestrian and cyclists" – "There are not enough bus shelters and it is hot to wait for the bus" – "A footpath on Waterford Road would assist walking and cycling. It is not a wide road and it is dangerous for students and other residents walking on it" – Footpaths on Boundary Road and Progress Road are not complete. They are overgrown in places with uneven surfaces – Young people push their bikes and walk through the bush track on the unformed section of Waterford Street to get to Gailes Station. Someone has formed an informal bridge over Sandy Creek by pushing a guardrail over the creek – There is no access off the Centenary Highway extension at Carole Park

(continued)

Table 5 (continued)

Transport theme	Frequency of mention (n = 47)	Selected quotes and observations
Personal safety	17% n = 8	– Fear of walking from Wacol Station because cars often flash their lights and make rude signs (particularly the young girls in the focus group) – "It does not feel safe walking on Sunday to get the train at Wacol Station as there are less cars and trucks and the factories are closed" – "There are tracks over the Logan Motorway where young people run across it to get to Gailes and Goodna"
Bus frequency, scheduling, reliability, convenience	15% n = 7	– The trains and buses do not synchronise as buses are always late. "The 8.25 am bus in the morning is always late to meet the train at Wacol Station. There isn't enough time for the drivers to get to the station. The timetable needs to be reviewed" – "You must allow plenty of time when travelling on the bus and train to get to appointments as the buses are often late, or they leave the train station before the train has arrived. You schedule your train travel to the bus as there are more trains and they are reliable. If you are leaving the city, you make sure you get the train that is scheduled to meet the bus at Wacol Station" – "Buses are late or not coming at all" (appointments missed)
Liveability	11% n = 5	– Cleanliness of bus shelters (wasp nests and spiders) – "We live in the ghetto of Brisbane" – "People don't like bus shelters because the kids sniff paint in them" – "Walking through the bush to Gailes Station makes your shoes dirty for going out"

(continued)

Table 5 (continued)

Transport theme	Frequency of mention (n = 47)	Selected quotes and observations
Travel time	9% n = 4	– "You need to leave on the bus an hour early to make sure you get the train to get to appointments on time" – "You get used to waiting" – It is about 45 min to Wacol Station and 25 min to Gailes Station if walking
Other (suburb identity, cooperation between transport service pro viders, the car is the most convenient)	9% n = 4	– No signs to indicate Carole Park at any access points to the suburb. "Need signs at Roxwell Street, Boundary/Progress Road corner, Wacol exit off Ipswich Motorway at least" – "Why do Forest Lake have nice bus shelters and Carole Park doesn't?" – There is a coordination and 'generosity' issue between Bris- bane Transport (BT) and Westside; for example, no Westside signage or timetables at the BT stop at Inala – "Car is the best way to get out of Carole Park to other areas"
Bus and train destinations	3% n = 2	– "No bus from Richlands State Primary School to Carole Park on Progress Rd in the afternoon. State School students need to walk to Richlands Tavern to get the bus home to Carole Park"

focus group are not confidential. The idea of the focus group is to maximise exchange between members. This means that confidentiality for group members is not possible within the group.

In my experience of focus groups in urban research, the more resources applied to planning and conducted the groups, the more effective they are in eliciting high quality research data. This means that the focus group method can be more resource intensive than alternatives such as online surveys or observation. Effective focus groups require planning and facilitation, as well as note taking or recording, time-keeping and post-group write-up and analysis. The resources and skills to conduct focus groups are a key reason for their limited use in urban research. It is hoped that new technologies such as online polls, simple interactive online geographic information systems and online meeting platforms are supporting focus groups to be more engaging, with less resources.

8   Conclusions

This chapter outlined the many contributions of the focus group method to urban research. The focus group is a commonly used qualitative research method that can leverage local community interest in a topic to elicit data for urban analysis. Focus groups can either be formed from existing interest groups or recruited just for the purposes of the research. Either way, the focus group is an adaptable method. From scoping and framing the research project to enriching understanding and prioritisation of issues and solutions, the focus group is a flexible method that adds value to most urban analysis, including quantitative-dominant research.

Focus groups complement many other urban research methods and they are most effective when they are deployed in mixed-methods research. The chapter described the many potential applications of the focus group and its association with other research methods including literature review, in-depth interviews, field work and observation and quantitative data analysis.

Key Points

- Focus group have multiple roles in urban research projects
- Focus groups must involve an exchange of information and ideas between focus group members
- Groups can be pre-existing or formed for the purposes of the research
- In urban research projects, focus groups are complementary to and enrich the results of other methods including literature review, stakeholder conversations, quantitative methods, in-depth interviews, field work, observation and larger group processes
- To be most effective, focus groups require skilled facilitation and preparation
- Technology such as online polls and other voting and prioritisation apps and online meeting platforms support focus groups to be less resource intensive and more engaging

- Ethics, confidentiality and resourcing are key issues for focus groups

Further information

For those wanting to get further information about focus groups, there are a number of useful guides. For specific material dealing with secondary data see:
- Richard Krueger and Mary-Ann Casey (2000). *Focus Groups: A Practical Guide for Applied Research*. Thousand Oaks: Sage Publications.
- Pranee Liamputtong (2019). Qualitative Research Methods. 5th edition. Oxford University Press.
- David Morgan (2019). Basic and Advanced Focus Groups. Sage Research Methods. Thousand Oaks: Sage Publications.

References

1. Morgan DL (2019) Basic and advanced focus groups. Sage research methods. Sage Publications, Thousand Oaks
2. Krueger RA, Casey MA (2000) Focus groups: a practical guide for applied research. Sage Publications, Thousand Oaks
3. Morgan DL (1988) Focus groups as qualitative research. Sage Publications, Newbury Park
4. Liamputtong P (2019) Qualitative research methods. 5th edition. Oxford University Press
5. Tong A, Sainsbury P, Craig J (2007) Consolidated Criteria for Reporting Qualitative Research (COREQ): a 32 item checklist for interviews and focus groups. Int J Qual Health Care 19(6):349–357
6. Johnson LM (2020) Transport justice in the suburbs: leveraging social capital to reduce transport disadvantage. PhD Thesis, School of Earth and Environmental Sciences, The University of Queensland

In-Depth Interviewing

Natalie Osborne and Deanna Grant-Smith

Abstract This chapter explores in-depth interviewing, a widely used method in qualitative research aimed at building a depth of understanding, rather than factual or abstract information. We give a brief account of structured interviews, but these have more in common with surveys and questionnaires than in-depth interviewing. Semi-structured interviews are interviews that follow a set of questions, allowing some flexibility in how questions are asked, and provide the researcher with the opportunity to ask clarifying and follow-up questions. Unstructured interviews are more flexible and open-ended. As such, they can be quite unpredictable and varied and are well suited to exploratory research and other research interested in meaning and experiences. There is a range of ethical considerations when using this method, including confidentiality establishing clear and reasonable expectations and offering appropriate reciprocity to research participants, and the potential of this method to cause harm, to both researcher and participants. In-depth interviewing is a powerful but demanding method for urban research; it is resource intensive and requires a lot of skills on the part of the researcher to design effective and ethical protocols and to generate meaning from complex, often messy, data.

1 Introduction

Urban research has traditionally privileged a technocratic approach to understanding urban life based on quantitative data and the aggregation of responses. By contrast, interviews are more interested in getting to the heart of personal experiences and privileging participant expertise and have become one of the most ubiquitous methods

N. Osborne (✉)
Griffith University, Nathan, QLD, Australia
e-mail: n.osborne@griffith.edu.au

D. Grant-Smith
Queensland University of Technology Brisbane, Brisbane, QLD, Australia

© Springer Nature Singapore Pte Ltd. 2021
S. Baum (ed.), *Methods in Urban Analysis*, Cities Research Series,
https://doi.org/10.1007/978-981-16-1677-8_7

in qualitative research [1]. Broadly, the research aims of in-depth, qualitative interviewing are to describe, understand, evaluate, analyse, and reflect [2] and the questions asked are designed to "elicit information that is 'factual', descriptive, thoughtful or emotional" [3, p. 107]. As a result, interviews do not tend to be used where the research goal is to test or prove a particular hypothesis or relationship, and, due to their emphasis on context, the goal of interviewing is not usually to produce knowledge that is generalisable. In-depth interviewing allows researchers to explore in detail people's subjective experiences, biography, and assumptions. Because of the value they place on experiential forms of knowledge, in-depth interviews are well suited to postpositivist and constructivist research, including (but not limited to) work informed by critical theory, and work taking a grounded theory approach [4, 5].

This chapter introduces different kinds of qualitative interviewing, outlining for each the kinds of questions they are best placed to answer, challenges, and ethical considerations, alongside useful tips and techniques. Throughout the chapter, empirical examples are provided to illustrate key points. We begin with a brief overview of structured interviews, noting that they are sometimes excluded from the category of qualitative or in-depth interviews as they bear more resemblance to surveys than other types of interviewing [6]. We devote more time to semi-structured and unstructured interviews, before moving on to their variants including online interviewing and group/multiparticipant interviews. Finally, we conclude with a discussion of some overarching issues and concerns related to in-depth interviewing.

2 Structured Interviews

Structured interviews are a staple in positivist research due to the tendency to focus on consistency in data collection and measurement, and "objectivity" in analysis. A structured interview uses an interview schedule with standardised questions, which are primarily close-ended, and participants typically choose between the same set of alternatives, which is why they are sometimes referred to as a "personally administered questionnaire" [7, p. 142]. Because structured interviews are so similar to surveys [8], it is not uncommon for them to be designed using the methodological assumptions and design principles of questionnaires rather than those of in-depth interviews. Researchers might choose a structured interview over a survey or self-complete questionnaire as they tend to have a higher response rate [7].

Unlike other qualitative interview approaches, there is no flexibility in the order that questions are asked, the way that they are worded, and there is no scope for adding new or follow-up questions. The level of participant–interviewer engagement is, therefore, low and there is typically no or little clarification of questions by the interviewer to reduce bias or variables [7]. They can be undertaken over the phone (e.g. political polls) or face-to-face, as in street intercept surveys, and by a team of interviewers (see box below for an example showing how this might be used in an urban planning context). These characteristics allow for a large sample size in a short period of time, and the application of a range of quantitative data analysis techniques. Although structured interviews are mainly geared towards producing quantitative data, they can also elicit qualitative data by using open-ended questions, but if you have a lot of open-ended questions, another interview technique might be more appropriate.

Determining public values of urban forests using street intercept surveys

Camilo Ordóñez with Peter Duinker, John Sinclair, Tom Beckley and Jaclyn Diduck [9] conducted street intercept surveys to determine what the public considers important or values about urban forests in Canada. Participants were recruited from the local streets of Fredericton (New Brunswick), Halifax (Nova Scotia), and Winnipeg (Manitoba). Four survey sites were chosen in each city based on a combination of high pedestrian traffic and the presence of treed spaces. Although the researchers adopted a non-selective recruitment of respondents, interviewers were instructed to achieve a target of at least 100 participants at each location. The survey contained a mix of demographic questions, Likert-scale questions on a 1–5 scale and questions requiring verbatim responses. A total of 1077 people participated in the survey, which took 3–5 min to complete. The researchers coded the verbatim responses so that they could conduct statistical analysis on the qualitative data alongside the quantitative data they had collected.

3 In-Depth Interviews

There are some key differences between semi-structured and unstructured interviews, but they are both "in-depth" interviews and have more in common with each other than they do structured interviews. We discuss some of these overarching commonalities first, before discussing each method in more detail.

Semi-structured and unstructured interviews are used in qualitative, exploratory research, research interested in lived experiences, local knowledges, the relationships between context and experience and/or actions, the creation of meaning for deepening our understanding of complex and/or poorly understood phenomena, and in research adopting an interpretive or critical approach to analysis. These methods are interested in understanding someone's experiences on their own terms, getting a sense of the context of a person's experiences, understanding their world and worldview, and how that might shape knowledge and meaning [1, 6, 8].

Knowledge is thought to emerge from encounter, and these encounters are relational, contextual, placed, embodied, and affective [10]. Knowledge comes from allowing oneself to be affected by the people, places, and/or relations being studied— to take up a position of wonder [11] and be moved. As such, people conducting in-depth interviews tend to believe that knowledge is not "discovered" by a researcher;

rather it is co-produced, a collaboration between participants, researcher, and other aspects of their encounter [12, 13].

Foregrounding emotion, bodies, and relationships may run counter to how we often think about research. Often we associate "doing research" with adopting a position of neutrality or objectivity, and taking a dispassionate approach. This is not always appropriate, and in fact, for the sorts of questions we might use unstructured interviews for, often it is not. Instead of seeking to escape or transcend our position, we instead critically reflect on our positionality, along with the positionality of our interview participant, and consider how they might interact and shape the research encounter, which is itself "rich with emotions and emotional dynamics" [10, p. 236].

Both types of interviews are typically, but not exclusively, conducted one-on-one and face-to-face, although there are exceptions on both counts (which we discuss later in this chapter). Both types ask mostly open-ended questions, with the intention of prompting a detailed and rich response. As responses tend to be longer and much less predictable than in structured interviews, in-depth interviewers will almost always ask, or perhaps require as a condition of participation, that the interview be audio-recorded. If the researcher relies on note-taking, not only will they miss content (particularly exact wording), it can be more difficult to maintain the conversational connection these types of interviews tend to aim for. Audio-recording for listening back, note taking, and/or full transcription, is standard practice for this kind of interview, though it is worth noting that some participants or potential participants may be deterred by this.

Unlike quantitative approaches that aim to choose a random or representative sample in order to create objective and replicate data, with in-depth qualitative interviews, the aim is not to be representative but rather to understand how individual participants experience, perceive, or make meaning of the phenomenon under investigation [3, 14]. As a result, there are very few rules to dictate sample size for interview-based research. It depends on the research question [15], diversity of the participant pool and the resources and capacity of both researcher and participants. There are instances where a very small number of interviews are sufficient, particularly for elite [16] or elusive, hidden or hard to reach populations [17, 18] or in autoethnographic self-interviewing, which recognises that we, ourselves, can be legitimate knowledge subjects [19]. It is, however, more common for in-depth interviews to include between 5 and 25 participants [20].

"Saturation" is often used to justify sample sizes in interview-based research (see box below). Monique Hennink, Bonnie Kaiser and Vincent Marconi [21, p. 591] argue that both code saturation (where the range of thematic issues are able to be identified) and meaning saturation (where a "richly textured understanding of issues" is developed) are important. Examining the transcripts of 25 in-depth interviews, they found that while code saturation was reached at nine interviews, the analysis of between 16 and 24 interviews was required to achieve meaning saturation. They suggest that "code saturation may indicate when researchers have 'heard it all', but meaning saturation is needed to 'understand it all'" [21, p. 591]. The openness of in-depth interviews may mean that it is difficult to know at the outset what your sample should be.

Some qualitative researchers internalise a kind of 'quantenvy', based on the misconception that more interviews are always better, or the inappropriate expectation that qualitative studies should mirror the large sample sizes common in quantitative research. This is perhaps especially the case in fields that use a variety of methods—like urban studies and planning—and/or where quantitative methods are preferred or privileged. In such contexts, the onus is often on the in-depth interviewer to justify their approach and demonstrate why it is most suitable for the research aims, and they might still be asked to justify their approach by the terms of quantitative research, like generalisability, validity, and reliability, that may be of limited relevance.

Importantly, *more data* are not always better when it comes to in-depth interviews. The richness of these data can overwhelm a researcher. Indeed, even with a modest participant pool, it is likely that there will be fertile seams of data that you would not have the time, capacity, or space to explore in your research. Further, as in-depth interviews are taxing for the researcher, pushing oneself to do too many can result in exhaustion, compromising your research. A burned-out researcher is likely to find it harder to stay focused, listen, empathise, and be a conscientious interlocutor.

Both semi-structured and unstructured interviews demand highly developed listening skills. Listening is a skill that is rarely actively taught, and yet it is one of the most important skills for an interviewer [1]. Despite the conversational dynamic, it is also important that the interviewer does not rush to fill a silence. A great interviewer is not only an attentive listener, they can create and hold space for their participant—a space that encourages deep reflection, contemplation, and honesty—and that may mean sitting in thoughtful silence.

Engaged listening—and being able to demonstrate that listening to the speaker/s—takes practice, and even when we are skilled at it, it can be tiring, painful, and distressing (see Sect. 3.5.3). It should always be affecting; listening requires an openness on the part of the listener, not just the speaker. The listening of the in-depth interview is not only attending to the words the participant is saying, but:

> "it also involves being attentive to non-verbal cues of listening and how we feel in our bodies and places of enquiry (Bissell, 2010). These non-verbal and embodied cues involve close cognisance of our emotions – distress, trauma, and sadness. Understanding feelings, emotions, and embodied atmospheres pertains to our role, as researchers, to listen in our respective field settings" [22, p. 19].

Here, Ratnam [22] reminds us also to listen to ourselves—knowledge emerges from the interview as an *encounter*, so we must attend to our own responses as well as that of the participant, as well as being mindful of the interpersonal and structural dynamics that sit between us.

"Saturation" in research with a hard-to-reach participant pool
Sutherland [23] employs saturation to guide her research with a hard-to-reach population—hobby farmers in a particular parish in Scotland, UK. Sutherland was interested in the land management practices of small, non-commercial

land holders, a group that can be quite difficult to reach and involve in social research. Saturation was used to determine how large her sample would need to be in order to be valid, and this was supported by a parish mapping approach. With the help of participants, a map of the parish was filled out identifying owners and tenants, which in turn was then used to guide the interview process. The "completion" of the map marked data saturation.

3.1 Semi-structured Interviews

Longhurst [3, p. 103] describes semi-structured interviews as "talking with people but in ways that are self-conscious, orderly and partially structured". This structure is provided through the use of an interview schedule or guide, which can contain a mix of prepared questions and themes for discussion. Semi-structured interviews are among the most common qualitative data collection techniques because they can accommodate a wide range of philosophical traditions. However, what binds those who chose semi-structured interviews is a commitment to surrendering some level of control to the participant in allowing them some latitude to guide the way that interview questions are approached and answered. Thus although the interviewer has a list of predetermined questions or themes, the order in which they are asked may follow the flow of the conversation, allowing participants to explore issues they feel are important and in an order that makes sense to them. In response, the semi-structured interviewer may ask some follow-up questions, which will not have been on the original interview schedule. As a result, there will be variability between interviews based on the participant's interests, experiences, and views [14].

It is important that semi-structured interviews do not ask close-ended or leading questions [24]. Having said this, it can be useful to commence the interview with questions that are more factual than abstract to ease the participant in, then once rapport has been established it becomes possible to ask more sensitive, difficult, or intimate questions [14]. Gill Valentine [14] suggests that starting questions with the phrase "tell me about…" can be an effective way of encouraging discussion as it is less inquisitorial and also reminds participants that you are interested in what they have to share. To get the most out of semi-structured interviews do not neglect the importance of sensitive probing, a question that is asked to follow-up and explore the issues raised during the interview. Probing provides the opportunity for the participant to provide additional information and reflections. Sometimes the issues that will need further probing can be anticipated in advance and potential probing questions listed underneath your main questions to remind you to probe further, particularly in your first few interview encounters [24].

It is important to anticipate there will be some differences not only between interviews but also between interviews conducted by different interviewers, as a result of the lower level of structure and greater degree of flexibility that chararacterises semi-structured interviews. Interpersonal dynamics also come more into play in this form of interviewing, which the interviewer needs to reflect carefully on, as these can influence results (see box below). In order to ensure that you cover all of the themes or key questions in your interview, it can be helpful to tick these from a list as you conduct the interview. We also keep brief notes as we conduct the interview to remind us of points we would like to follow-up without interrupting the flow of the interview. Because this means that there will be a lot of information that you will need to monitor, when scheduling your semi-structured interviews be mindful of how many you try to fit into a day. Although it is not always possible we try to limit to no more than three semi-structured interviews a day.

Understanding working women's transportation needs and experiences in Mexico City through semi-structured interviews

Lucia Mejia-Dorantes' [25] exploratory study used in-depth semi-structured interviews to elicit information from women working or living around a focal point in Mexico City. The aim of the research was to better understand the transport and mobility characteristics of medium to low-income working women in a metropolitan area and to understand the main factors that influence their travel patterns, how the different services are perceived and evaluated, and their daily travel constraints.

The researcher conducted 22 interviews with women who were selected through opportunity sampling and snowball recruitment. The interviews were audio-recorded. Participants were asked questions about the different transport systems in the Metropolitan area, how they used them and the changes they have perceived and other problems related to public and private transport. A short survey was also filled-in during the interview to gather demographic information.

Reflecting on the data collection process, Meija-Dorantes argues that participants were likely to share personal experiences with her because she was a local woman who had the time and willingness to hear their point of view about their problems and experiences and needs.

3.2 Unstructured Interviews

Research employing unstructured interviews may come from a range of philosophical traditions but are generally affiliated with constructivist and interpretive research, as they are designed to "elicit people's social realities" [8, p. 239]. There are other papers that explore in more depth the philosophical alignments of different interview techniques (see 2; 1], but here we will consider two key assumptions about knowledge unstructured interviews reflect. First, the selection of this interview approach suggests a strong belief that context, positionality, and relationships are central to understanding the phenomena being studied, and that knowledge about those phenomena *cannot be abstracted* from the context in which it was produced, or from the people who hold and create that knowledge. In simple terms: positionality matters, people matter. Second, unstructured interviews reflect the belief that people can be experts on their own experiences, that non-academics create knowledge and engage in analysis, and that we can learn from how people draw meaning from their experiences. In short, the identity and positionality of both researcher and participant *matters* in unstructured interviews, and the knowledge produced will vary based on who did the interview [26]; this is not something to be avoided or minimised, but something to be reflected on and drawn into the analysis.

Unstructured interviews offer greater scope for depth than other types of interview techniques; they also tend to be the most time-consuming way to conduct interviews, and data analysis is more complex. With structured or semi-structured interviews, there are threads of commonality across interviews, and it can be relatively straightforward to compare and contrast responses, and so derive insights. The variation in unstructured interviews can make it difficult to see them as a dataset.

In unstructured interviews, the researcher may have a loose list of prompts, or topics they would like to explore with the participant, or goals for the interview, but the format is loose, open-ended, and flexible [2]. These interviews tend to be more conversational, and more of a dialogue or social encounter between researcher and participant [27], and the questions the interviewer asks will be guided primarily by the participant's responses [8]. The participant has more autonomy and greater ability to shape, direct, and possibly even lead the interview than in other forms of interviewing, and in some cases, it may be more fair to think of participants as co-creators—they are shaping the interview, and indeed the research, at least as much as the interviewer.

Unstructured interviews can be conducted in a number of different ways. Avoid scheduling unstructured interviews back-to-back; the length of the interview itself is unpredictable, and, depending on the subject matter covered, you might find you need more time to process and reflect before you are ready for another go. Because of this, unstructured interviews might not be an appropriate technique if you have a very constrained or limited time for fieldwork, unless you only need a very few.

When organising the interview, you can adopt the standard "appointment" approach, where you arrange a time and place to meet with your participant. More organic approaches may be appropriate too—as long as your ethical protocols are followed, you might find that unstructured interviews work well while engaged in a shared activity; e.g. interviewing activists at a blockade, interviewing volunteers as you undertake dune revegetation, shadowing someone at their workplace and interviewing them at quiet times during the day, or conducting your interview as the participant takes you on a walk around their neighbourhood (or another place relevant to your study). Unstructured interviews are often conducted as part of a broader methodology that includes ethnographic and/or participant or participatory observation [8], so this more organic, blended approach may be a better fit.

Unstructured interviews may require a "warm up" period—a time spent with small talk, the researcher and participant getting comfortable with each other, before the formal interview commences [28]. This might be a chat immediately preceding the interview, or it might be one or more separate meetings/conversations before the interview is even scheduled. In some cases, especially where the participant pool has a heightened need for privacy, you might need quite a long period of relationship building before your interviews can occur. The degree of preamble will depend on the context of the research, the sensitivity of the topic, the guardedness or wariness of participants, whether or not the researcher is already known to participants, and more. Trust and depth may also be aided by interviewing the same participant repeatedly over the course of the project. Relationship building supports the intimacy and vulnerability of unstructured interview encounters (see box below)—and of all kinds of data collection, unstructured interviews are perhaps the most intimate, and may demand or produce the greatest vulnerability.

Because they are more open-ended, it is possible that the interview will go to unanticipated places; while that is a large part of why we might use this method, you may not have prepared for the information you receive. For instance: a participant might share their participation in illegal activity—how would you navigate that? You might inadvertently touch on a traumatic experience, and the participant could become very distressed. You might be the first attentive listener the participant has spoken to in a while, and they might open up about a personal issue that is not particularly relevant to the research. The participant might get carried away and then later regret the things they said. These things may happen in semi-structured interviews too, they are just more likely in the more unstructured forms of interviewing because the interviewer has relinquished more control.

It can also be difficult to navigate boundaries as you adopt a more conversational and less formal approach. Depending on the topic and/or the participant, you may find it appropriate to share a bit more about yourself, or openly reject a detached or "neutral" stance. And yet you do not want to take up too much space, or centre yourself in the process. How you draw and maintain boundaries with research participants will depend on your research topic, who they are, who you are, if you had a prior relationship, how long the research relationship is, and more. A great unstructured interview can feel like a profound connection—"ghosting" a participant afterwards might leave them feeling exploited, or silly. You may want to check in with participants after the interview and have follow-up chats. You could also offer participants additional opportunities to be involved in the research process, whether that is in reviewing their transcript, contributing to data analysis, and providing them with copies of your work. Navigating these relationships responsibly is not always easy, and it is important for the researcher to be able to access advice, process, and debrief with someone who is *not* a participant.

Analysing unstructured interview data is complex and time-consuming, and may require the researcher to take an iterative approach and revisit the data several times. It is quite likely that the interviews might seem all quite different from one another at first, and it can be difficult to find "the story". Thematic analysis is perhaps the most common approach to analysis, and this might include a mix of inductive, deductive, and axial coding. It tends to be at this point where weaknesses or gaps in the research aims and/or theoretical framing causes trouble, and you might find you have little to hang on to or guide you through a complex and contradictory mass of data. It may take several rounds of immersing yourself in the data, then taking a step back to try and discern the broader patterns and themes, perhaps returning to your aims, theory, and the literature, before you find the story/stories your data tell.

> **Understanding unconcern about climate change using unstructured interviews**
> Lucas and Davison [29] used in-depth, unstructured interviews to better understand attitudes of unconcern about climate change in Australia. Rather than assuming all those unconcerned about climate change are victims of propaganda or bias, the aim of this project was to explore the social relationships of unconcern, and to look at unconcern with an empathetic gaze in the context of other concerns, values, lived experience and material realities. Lucas interviewed seven participants eight times in a 6-month period. Returning to participants over and over allowed her to build trust and an understanding of their social worlds before even broaching the topic of climate change (which she did not bring up until the fifth interview with each participant). This commitment to building trust, demonstrating respect, and taking participants' lives and concerns seriously, allowed the researchers to identify a diverse set of dynamics and contextual factors shaping unconcern about climate change, with important implications for climate change communication and policy.

3.3   Other In-Depth Interview Formats

In addition to the broad categories of in-depth interviews described above, there are a number of variants. Due to space limits, we will confine ourselves to two—using digital technology for interviews and group interviewing.

3.3.1   In-Depth Email Interviewing and Other Digital Approaches

Although face-to-face remains the generally preferred medium for interviews and even considered by some as the "'gold standard' of qualitative research" [30, p. 613], there are other options that may, in some cases, be appropriate or even preferable. Non-face-to-face interviews—e.g. by telephone, video call, and email—can broaden the participant pool by including people across distance and timezones and by being more accessible to people who might otherwise face difficulties participating face-to-face, e.g. due to caring responsibilities, disability, illness, shift work [31]. They can also be more accessible for the researcher for the same reasons and can help researchers navigate political or health-based travel bans. Further, they reduce the economic and environmental costs of travel, which is an important consideration. However, as with all data collection methods, their use must be justified on both practical and methodological grounds and attention must be paid to the way that the technologies may shape or alter the interview encounter [30].

Audio-only interviews (via telephone) have been used successfully for a long time, and now video calling tools like Skype and Zoom are becoming common place [32]. Adams-Hutcheson and Longhurst [33, p. 148] argue that while online interviews may "feel different" to offline interviews neither is inherently better or worse than the other. Jenner and Myers [34] found conducting interviews online had little impact on their ability to build rapport, interview length, cancellations, or on the willingness of participants to share deeply personal experiences. In fact, the distance offered by online interviews may be well suited to interviews dealing with sensitive subject matter [35]. However, the interview may feel less formal, and it may be easier for the participant to forget it is being recorded, which Weller [30] cautions may lead to them divulging more than they would, on reflection, wish. Further, familiarity with technology (or indeed, access to technology) should not be taken for granted and the researcher may need to offer directions, training, or support [35, 36]. Researchers do well to choose a technology that is free, and one that is likely to be familiar to participants. Researchers should ensure they use a professional rather than personal account, and seek training or support on how to handle inappropriate or unwanted contact, including unrelated sexually explicit communication or images, particularly when interviews are conducted anonymously [35].

Asynchronous online interviewing is also an option. In-depth email interviewing, by a series of back-and-forth emails over a given period of time, has the potential to elicit deeply reflective answers without the need for face to face interaction [37, 38]. Email interviews can be semi-structured or unstructured, starting with a single question as a prompt for a more organic exchange [39]. Like any form of technologically mediated interview, email interviewing reduces the need to travel, and transcription is not required, making it very cost-effective [38]. Although the back-and-forth takes time, this extended approach allows both researcher and participant time for iterative reflection [40]. Ratislavová and Ratislav [38] argue that email interviews lower the barrier to participation by allowing participants to participate in their own home, and in their own time. This high level of participant control can be especially important in research on sensitive topics (see box below).

Data safety and management are additional concerns. Email systems should be password protected and not able to be accessed by others [31], and the interview should be downloaded in a way that preserves the order of the exchange [38], and the emails themselves deleted on completion [31].

This approach may still be mentally and emotionally demanding for the researcher. Fritz and Vandermause [37] suggest that in order to manage different email conversations, only three should be conducted concurrently, and that for the most authentic responses researchers must be flexible and respond to different participants' rhythm, flow, and preferences. The researcher still needs to establish rapport, ask appropriate questions, actively listen and end the email interview appropriately [40]. Emoticons can help communicate tone and simulate non-verbal cues [37, 38].

> **Understanding how millennials make sense of their lived experiences in mixed-use communities using in-depth email interviews**
> Bowden and Galindo-Gonzalez [40] used in-depth email interviews to conduct an exploratory pilot study to understand millennials' experiences of mixed-use neighbourhoods. This method was well suited to their prospective participant pool as millennials value flexibility and are comfortable communicating online. Establishing relationships and rapport with participants was still a prerequisite to gathering rich data. They observed that embedding questions within the body of an email results in a significantly better response rate than providing them in an attachment, and that they should not ask too many questions at once.

3.3.2 Group/Multi-participant Interviews

While we have primarily discussed interviews as comprising a researcher and one participant, an interview can also have multiple participants. Group interviews, or multi-participant interviews, are distinct from a focus group. In a focus group, the emphasis is on the interaction between participants, and the interviewer might be relatively passive, while in an interview, the focus is on the interaction between interviewer and participant [3]. This also holds true for multi-participant interviews; while participants will interact with each other, the interviewer remains central to the encounter.

Group interviews are any interview where there is more than one participant. They are often favoured for interviewing children [41], and may otherwise be useful when participants are hesitant or in need of someone to support them in their participation (see box below). Often group interviews are planned and instigated by the researcher, but sometimes the participant might request it themselves. You may need to allocate more time for a group interview, and you are likely to hear less from each participant than you would if you were one-on-one. Some participants can dominate the discussion, which may need to be managed by the interviewer. The informed consent process can be more complex too, and participants may, inadvertently or otherwise, breach the confidentiality of other participants [42].

Group interviews can become very conversational. Whatever control the interviewer is surrendering—whether they are taking a semi-structured or unstructured approach—is then shared by all participants, which can be generative, but may also heighten the potential for conflict. Where participants are well known to each other and the encounter flows as a conversation, the interviewer may find they begin referencing events and people or matters well known to other participants, forgetting that you do not know the backstory, so you may need to interject more. These interviews can also be time-consuming and/or costly to transcribe; each speaker including the interviewer needs to be identified in the transcript, and there is a higher likelihood of overlaps in talk, which can be hard or impossible to capture in the transcript.

> **Unanticipated group interview when studying emotional geographies of mine closure**
> Osborne [43] was conducting unstructured, storytelling-based interviews in her research, and she had arranged to meet a young woman at a cafe for an interview. She was a little nervous and had brought a friend along, not as passive support, but to contribute. After a chat, the friend called a family member who she thought would have lots to say, who promptly joined them at the cafe. It was a long, rich group interview between participants all well known to each other and enabled people to participate who otherwise might not have felt comfortable. Each participant had to be taken through the ethics protocols and sign a consent form, so this is a good reason to carry spares!

3.4 Other Considerations in In-Depth Interviewing

In this final section, we consider a range of contemporary considerations in interviewing including how technology is shaping interview practice and ethical concerns like confidentiality, reciprocity, the potential for harm in interviewing.

3.4.1 Confidentiality and Anonymity

Anonymity is one of the primary ethical considerations in human research. It is often standard to offer anonymity to participants, but this is rarely guaranteed. Even if you redact their name, if you refer to other aspects of their identity that might be relevant to your project (e.g. gender, age, sexuality, race, ethnicity, dis/ability, their affiliation with particular groups, the type of work they do, their family type, that they live in a particular area), it may become easy for people to infer who a participant is. De-identification is most effective in large datasets and participant pools, and where no one datapoint is given particular attention. But in-depth interview data do not tend to aggregate that way; it is often counterproductive to abstract data from the participant who generated it.

It is important to be honest with your participants about the limits of anonymity in your particular research project, so that they can make an informed decision about if and how they will participate. It is common practice, for instance, to refer to participants by a descriptive code (e.g. "pedestrian F20s" might refer to a woman in her mid 20s who walks to work), or by a pseudonym. Selecting a pseudonym can be more complex than it seems—should the pseudonym preserve the ethnicity of names (e.g. Amy becomes Mary), or might that compromise the participant's anonymity (particularly if they belong to a minority ethnic group in a fairly homogeneous participant pool)? You may wish to ask your participants to select their own pseudonym in discussion with you—while also reminding them that it is always still possible that a reader might deduce who they are. Some participants may be unconcerned, or prefer you to use their name. For instance, in Osborne's [43] Ph.D. research, an Aboriginal elder being interviewed, Aunty Donna Ruska, insisted her name be used. She said her people were given few opportunities to have their stories heard and she did not want her story stripped from her identity.

When reporting your research, avoid quoting or making direct reference to things that might identify participants. It is not always clear what details are identifying or not, so you might offer to let your participants review their interview transcript (or any direct quotes you wish to use) and redact anything they wish. This is also good practice if your interview covered sensitive topics; sometimes people tell you things because they want you to know them, but they do not want you to directly report on them.

Finally, consider fictionalising the data by using it to tell a story that unfolds in a fictional place [44]. This approach is common in business studies or education where companies or schools are fictionalised, however, in urban and geographical research place is usually one of our primary concerns, so displacing data might be counterproductive. Another approach would be to combine the stories of several participants to form "characters" or "archetypes", and report your findings through those amalgams [45]. Concerns with this approach are that it could essentialise or stereotype participants, muddle important contextual factors, and/or participants may resent not being to identify themselves in your work, or feel misrepresented if they do not identify well with how you have compiled your amalgams. Satchwell [46] actively involved their participants in creating the fictionalised accounts, which may avoid some of these issues while offering unique insights, but this could not be achieved through interviews alone, and additional methods would need to be employed.

3.4.2 Expectations and Reciprocity

Some communities, especially marginalised communities, do not trust researchers, and may not welcome a researcher's scrutiny; outsider interest can be dangerous [47]. If you are interested in conducting research with communities that have been harmed by researchers in the past, and/or who continue to experience oppression and exploitation, you may need to adopt additional ethical protocols and design your methodology with this in mind, starting by asking if you are the right person to do this research. How will you minimise harm, and how can you give back?

Aboriginal and Torres Strait Islander People have by and large been overresearched, and yet that research has generally brought them little benefit [48, 49]. Often, researchers have conducted what Martin [49, p. 203] calls *"terra nullius research"*, which reinscribes the erasure and violence of settler colonialism. As such, some institutions have created detailed protocols particularly for doing research with Indigenous People. These offer some guidance but do not necessarily reflect the complexity of doing research with Indigenous Peoples, or the need to consider particular contexts and positionalities.

Researchers are often encouraged or required by ethics protocols to recruit participants via "gatekeeper" organisations and groups, especially when those participants belong to marginalised groups. While this may be appropriate, Sullivan [47] found that these organisations may silence the people they are ostensibly seeking to represent and/or protect. She maintains that these organisations are important and should not be circumvented, but that our engagement with them should be reflexive, critical, thoughtful.

Institutional guidelines may be little more than an "an exercise in compliance and risk management" [47, pp. 9–10]; it takes more than that to do research ethically. In-depth interviews are relational and situated; ethics too are about relations and relationships [49, 50]. Just as the knowledge produced by in-depth interviews cannot be abstracted, nor should the knowledge we seek in research be abstracted from considerations about the ethical implications of knowing, of claiming to know, and what knowledge might be used for.

Reciprocity is an important consideration, especially in research with marginalised people or communities keenly affected by an issue. It is beyond the scope of this chapter to provide a full discussion of reciprocity, but it is important to reflect on how the method of in-depth interviewing may increase participants' expectations—reasonably or otherwise—for reciprocity. In-depth interviews often rely on relationship building, and they ask a lot from participants (e.g. time, vulnerability, sacrificing privacy). Participants may come to see you as a trusted expert, confidant, or friend, and that relationship might increase the expectation that they should receive some benefit from having offered you their time, knowledge, and/or access to their networks.

Finding meaningful practices of reciprocity in research is important—ideally, these should be developed in conversation with your participants and the affected community, and should reflect their needs, your capacities, and the kind of work you are doing. Examples include:

- Giving a seminar or workshop to share your findings
- Providing feedback to relevant organisations in a manner they prefer
- Joining advisory committees
- Assisting with the writing of policy/submissions
- Offering training/skill development
- Showing up to, and supporting, a community's events and causes.

What reciprocity entails will vary, but it is also important to manage expectations about what you are capable of. Sometimes researchers are imagined to have more power and influence than they actually do—participants may agree to an interview because they think the researcher will be able to intervene on their behalf, but typically our influence is far less direct. In-depth interviewing can intensify this, and you may increase social pressure and your sense of obligation than if you were more distant from participants. Be clear about what is possible for you to do, do not inflate your importance in the hopes of recruiting participants, and remember that over-promising and under delivering can destroy trust and relationships, and make it difficult for you or others to do research with that particular group again.

3.4.3   Interviewing and Harm

In-depth interviewing may cause harm, particularly if the subject matter is sensitive, traumatic, controversial, or otherwise difficult to talk about. By asking participants to share their experiences, stories, and opinions with us, we may be asking them to relate deeply painful matters, and revisit grief and trauma. Even when the subject matter is not particularly sensitive, the more room a researcher gives their participants to explore a topic, the greater the likelihood that they might touch unexpectedly on topics that are painful for the participant (or, indeed, for the researcher).

In-depth interviewers should consider the likelihood that the interview may cause distress, and create a plan for minimising harm. This might include being sensitive to timing (not rushing in after a loss/disaster), providing information on counselling services, and reminding participants of their agency in the process—that what and if they share is up to them, that they can refuse, and that they can pause or stop the interview at any time. It may also be appropriate to check in with participants after interviews (e.g. the following day or week) to see how they are going and offer support then if needed. Most qualitative researchers are not trained counsellors, so it is important not to step into that role, but we have a responsibility to extend our consideration to participants, and consider what is ethically and interpersonally required of us when we ask for people's participation in our project.

That said, many participants report that participating in in-depth interviews, even on distressing or traumatic subject matter, can be positive, even healing [28]. Taking the time and space to talk about oneself to an attentive listener, a listener who is taking what is shared seriously, honouring it, and building it into something bigger, can be cathartic, even empowering [51, 52]. This underscores the importance of reporting back and/or sharing findings and publications—it demonstrates to participants that their contributions, even if they were painful to share, were taken seriously and valued, and that their involvement may help others.

We have argued throughout that in-depth interviews are *relational encounters* between researcher and participants, and that it matters who is in the room (whether literally or digitally). But there are power asymmetries in interview relationships, and potential for harm, hostility, dislike, and worse. We should not assume that research interviews are inherently "warm, caring, and empowering dialogues" [53, p. 480], even though that might be the ideal. It is imperative that researchers are sensitive to power and interpersonal dynamics and the impact they can have both on the data generated and interview participants' experience of the interview process. Further, although it is relatively common for researchers, and for institutional ethics procedures, to consider the potential harm or distress a participant might experience as a result of an interview, insufficient attention is paid to the emotional and psychological impact conducting interviews may have on the interviewer [22]. It is quite possible for the researcher to find themselves affected by the interview, in ways that also might be distressing, uncomfortable, and/or traumatic. One way of minimising this for both researcher and participant is to follow Valentine's [14] suggestion to "warm down" the interview as it draws to a close by asking more relaxed or light-hearted questions that the interview interaction ends on a somewhat positive note.

There are also more direct forms of harm a researcher may experience, and some of the practices of in-depth interviewing—emphasising relationality and openness, muddling boundaries between researcher and participants, using gatekeepers and key informants, spending lots of time alone with participants, possibly over a number of days/visits—can increase exposure to harm. Sexual assault, harassment, and other forms of violence are not uncommon for researchers doing interview-based field-work. Some researchers face heightened risks due to aspects of their identity and/or position (e.g. their gender, sexual identity, race, age, lack of seniority etc.), and research institutions and academic communities have often ignored or externalised these risks [54, 55]. While it is essential for in-depth interviewers to be aware of and critically reflect on power dynamics and positionality in their research, it is also crucial not to flatten out these relationships, or assume that either power or vulnerability to harm is fixed exclusively with one party or another.

4   Conclusions

In disciplines that privilege quantitative research, in-depth interviews may seem like a soft option. They are anything but. They require a high level of commitment and skill on the part of the researcher and can involve a significant degree of ethical, theoretical, methodological, and analytical complexity. They do have drawbacks; they are time-consuming and often resource-intensive, they can include physical, emotional, and/or psychological risks to both interviewer and participants. Depth can come at the cost of breadth, the data can be very difficult to work with, and the results may have limited general applicability. Further, as the interviewer often works alone, it can be difficult to access the kind of training, mentoring, and support, and receive the feedback, that helps us become great interviewers (see box below for other helpful resources!).

However, the "pros" of this method is that the depth of understanding, nuance, and context they can offer are rich and unparalleled. Further, interviewing honours the lived experiences of participants. Interviewing is not inherently feminist but it has been the work of feminists to elevate and legitimise this approach to research, arguing that participants and researchers offer different but "equally important components of the knowledge project" [19, p. 146].

Often, urban research has neglected and excluded diverse ways of knowing and the people who hold them [56]. However, for some research aims, the identities, positions, contexts, and situations of both researcher and researched are not only important, they are essential to the work being done. Instead of trying to adopt a "view from nowhere", researchers can actively seek out people with particular perspectives, situated in particular places and sets of relations, and draw on their own positionality to inform and enable their research [57]. In-depth interviewing provides a method to facilitate depth of understanding, and some questions we ask in urban analysis demand just that.

Key Points

- In-depth interviews allow us to explore complex, situated, uncertain, and unfolding issues with depth and nuance. They can help us learn how people understand issues and their worlds, make meaning out of events, and relate to each other and to place.
- In-depth interviews can be an effective and meaningful way to engage with marginalised peoples, communities, and knowledges.
- In-depth interviews can be semi-structured or unstructured and range from quite formal to very informal. The interviewer may use techniques that draw from story-telling, oral histories, and ethnographic research, and may speak to participants just once, or many times over the course of a project.
- In-depth interviews typically happen face to face and are audio-recorded, however they can also be conducted by phone, video call, or even over letters or email.
- In-depth interviews are very resource-intensive, and the data they produce can be uncertain, highly complicated, and difficult to work with.
- While in-depth interviews are often used in emancipatory research, or research otherwise concerned with questions of power and social justice, there are physical, ethical, psychological, and emotional risks to both researcher and participant that need to be carefully considered.

Further information

For those wanting to get further information about conducting in-depth interviews, there are a number of useful guides.

- MacCallum et al. [58] *Doing research in urban and regional planning: Lessons in practical methods,* Routledge offer a great handbook on urban planning research, which can provide more guidance on how interviews fit within a broader research process.
- Knapik [59]. The Qualitative Research Interview: Participants' Responsive Participation in Knowledge Making. International *Journal of Qualitative Methods,* 77–93 offers a thoughtful, reflexive account of participants' reflecting on the experience of being interviewed—this account of how participants may experience the research encounter can help us reflect on our practices.
- Braun and Clarke [60] One size fits all? What counts as quality practice in (reflexive) thematic analysis?, *Qualitative Research in Psychology,* provide

additional reflection and guidance around thematic analysis, one of the most common ways interview data is analysed.

References

1. Qu SQ, Dumay J (2011) The qualitative research interview. Qual Res Account Manag 8(3):238–264
2. Fossey E, Harvey C, Mcdermott F, Davidson L (2002) Understanding and evaluating qualitative research. Aust N Z J Psychiatry 36(6):717–732
3. Longhurst R (2010) Semi-structured interviews and focus groups. In: Clifford N, Cope M, Guillespie T, French S (eds) Key Methods Geogr, 3rd edn. SAGE, London, pp 143–156
4. Charmaz K, Belgrave LL (2012) Qualitative interviewing and grounded theory analysis. In Gubrium, JG, Holstein JA, Marvasti AB, McKinney KD (eds) The SAGE handbook of interview research: the complexity of the craft. SAGE
5. Given LM (2008) The SAGE encyclopedia of qualitative research methods. SAGE, Thousand Oaks
6. Kendall L (2008) The conduct of qualitative interviews: research questions, methodological issues, and researching online. In: Coiro J, Knobel M, Lankshear C, Leu DJ (eds) Handbook of research on new literacies. Routledge, New York, pp 133–149
7. Bougie R, Sekaran U (2020) Research methods for business: a skills building approach. Wiley
8. Zhang Y, Wildemuth BM (2017) Unstructured interviews. In: Wildemuth B (ed) Applications of social research methods to questions in information and library science, 2nd edn. Libraries Unlimited, Westport, pp 239–247
9. Ordóñez C, Duinker PN, Sinclair AJ, Beckley T, Diduck J (2016) Determining public values of urban forests using a sidewalk interception survey in Fredericton, Halifax, and Winnipeg Canada. Arboric Urban Forestry 42(1):46–57
10. Bondi L (2005) The place of emotions in research: From partitioning emotion and reason to the emotional dynamics of research relationships. In: Davidson J, Bondi L, Smith M (eds) Emotional geographies. Ashgate, Hampshire, UK, pp 231–246
11. Ahmed S (2004) The cultural politics of emotion. Routledge, New York
12. Anderson JM (1991) Reflexivity in fieldwork: Toward a feminist epistemology. Image: J Nurs Scholarship 23(2):115–118
13. Finlay L (2002) Negotiating the swamp: the opportunity and challenge of reflexivity in research practice. Qualit Res 2(2):209–230
14. Valentine G (2005) Tell me about… using interviews as a research methodology. In Flowerdew R, Martin D (eds) Methods in human geography: a guide for students doing a research project (2nd ed). Addison Wesley Longman, Edinburgh Gate, pp 110–127
15. Baker SE, Edwards E (2012) How many qualitative interviews is enough? Expert voices and early career reflections on sampling and cases in qualitative research. National Centre for Research Methods Review Paper. Economic and Social Research Council
16. Odendahl T, Shaw AM (2011) Interviewing elites. In Gubrium JF, Holstein JA (eds) Handbook of interview research. SAGE, pp 299–316
17. Lambert EY (1990) The collection and interpretation of data from hidden populations. NIDA Research Monograph 98. US Department of Health and Human Services, National Institute of Drug Abuse, Rockville MD
18. Sydor A (2013) Conducting research into hidden or hard-to-reach populations. Nurse Research 20(3):33–37

19. Crawley SL (2012) Autoethnography as feminist self-interview. In: Gubrium JF, Holstein JA, Marvasti AB, McKinney KD (eds) The SAGE handbook of interview research: The complexity of the craft. SAGE, pp 143–161
20. Creswell JW (2007) Qualitative inquiry and research design: Choosing among five approaches, 2nd edn. SAGE, Thousand Oaks
21. Hennink MM, Kaiser BN, Marconi VC (2017) Code saturation verses meaning saturation: how many interviews are enough? Qual Health Res 27(4):591–608
22. Ratnam C (2019) Listening to difficult stories: Listening as a research methodology. Emot Space Soc 31:18–25
23. Sutherland L-A (2020) Finding 'Hobby' Farmers: A 'Parish Study' methodology for qualitative research. Sociologia Ruralis 60:129–150
24. Healey-Etten V, Sharpe S (2010) Teaching beginning undergraduates how to do an in-depth interview: A teaching note with 12 handy tips. Teach Sociol 38(2):157–165
25. Mejia-Dorantes L (2018) An example of working women in Mexico City: How can their vision reshape transport policy? Transp Res Part A: Policy Pract 116:97–111
26. Muhammad M, Wallerstein N, Sussman AL, Avila M, Belone L, Duran B (2015) Reflections on researcher identity and power: the impact of positionality on community based participatory research (CBPR) processes and outcomes. Critical Sociol 41(7–8):1045–1063
27. Roulston K, Choi M (2018) Qualitative interviews. In: Flick U (ed) The SAGE handbook of qualitative data collection. SAGE, London, pp 233–249
28. Corbin J, Morse JM (2003) The unstructured interactive interview: Issues of reciprocity and risks when dealing with sensitive topics. Qualitat Inquiry 9(3):335–354
29. Lucas CH, Davison A (2019) Not 'getting on the bandwagon': When climate change is a matter of unconcern. Environ Plann E: Nat Space 2(1):129–149
30. Weller S (2017) Using internet video calls in qualitative (longitudinal) interviews: some implications for rapport. Int J Soc Res Methodol 20(6):613–625
31. Ison NL (2009) Having their say: email interviews for research data collection with people who have verbal communication impairment. Int J Soc Res Methodol 12(2):161–172
32. Cachia M, Millward L (2011) The telephone medium and semi-structured interviews: a complementary fit. Qualit Res Organ Manag 6(3):265–277
33. Adams-Hutcheson G, Longhurst R (2017) 'At least in person there would have been a cup of tea': interviewing via Skype. Area 49(2):148–155
34. Jenner BM, Myers KC (2019) Intimacy, rapport, and exceptional disclosure: a comparison of in-person and mediated interview contexts. Int J Soc Res Methodol 22(2):165–177
35. Sipes JB, Roberts LD, Mullan B (2019) Voice-only Skype for use in researching sensitive topics: a research note. Qualit Res Psychol 1–17. Online ahead of print
36. Mirick R, Wladkowski S (2019) Skype in qualitative interviews: participant and researcher perspectives. Qualit Report 24(12):3061–3072
37. Fritz RL, Vandermause R (2018) Data collection via in-depth email interviewing: lessons from the field. Qual Health Res 28(1):1640–1649
38. Ratislavová K, Ratislav K (2014) Asynchronous email interview as a qualitative research method in the humanities. Human Affairs 24:452–460
39. Nehls K (2013) Methodological considerations of qualitative email interviews. In: Sappleton N (ed) Advancing research methods with new technologies. IGI Global, pp 303–315
40. Bowden C, Galindo-Gonzalez S (2015) Interviewing when you're not face-to-face: the use of email interviews in a phenomenological study. Int J Doctoral Stud 10(12):79–92
41. Lewis A (1992) Group child interviews as a research tool. British Educ Res J 18(4):413–421
42. Valentine G (1999) Doing household research: interviewing couples together and apart. Area 31(1):67–74
43. Osborne N (2014) Stories of Stradbroke: emotional geographies of an island in transition. (Doctor of Philosophy), Griffith University, Queensland, Australia
44. Campbell A, Kane I (2012) School-based teacher education: telling tales from a fictional primary school. Routledge

45. Nevin J, Campbell A (2005) Fictionalising research data: towards a typology of issues for interviewing women in management positions. Paper presented at the British educational research association annual conference, University of Glamorgan, 14–17 September http://www.leeds.ac.uk/educol/documents/153953.htm
46. Satchwell C (2019) Fictionalised stories co-produced with disadvantaged children and young people: Uses with professionals. In: Jarvis C, Gouthro P (eds) Professional education with fiction media. Palgrave Macmillan, Cham, pp 49–69
47. Sullivan CT (2020) Who holds the key? Negotiating gatekeepers, community politics, and the "right" to research in Indigenous spaces. Geogr Res. Online ahead of print
48. Knight JA, Comino EJ, Harris E, Jackson-Pulver L (2009) Indigenous research: a commitment to walking the talk. The Gudaga study—an Australian case study. Bioethical Inquiry 6(4):467–476
49. Martin KBM (2003) Ways of knowing, being and doing: A theoretical framework and methods for Indigenous and Indigenist research. J Australian Stud 27(76):203–214
50. Bawaka Country, Wright S, Suchet-Pearson S, Lloyd K, Burarrwanga L, Ganambarr R, Ganambarr-Stubbs M, Ganambarr B, Maymuru D, Sweeney J (2016) Co-becoming Bawaka: towards a relational understanding of place/space. Prog Hum Geogr 40(4):455–475
51. Hooks B (2010) Teaching critical thinking: practical wisdom. Routledge, New York
52. Sandercock L (2001) Out of the closet: the importance of stories and storytelling in planning practice. Planning Theory and Practice 4(1):11–28
53. Kvale S (2006) Dominance through interviews and dialogue. Qualitative Inquiry, 12(3): 480-500
54. Caretta MA, Jokinen JC (2017) Conflating privilege and vulnerability: a reflexive analysis of emotions and positionality in postgraduate fieldwork. Profess Geogr 69(2):275–283
55. Ross K (2015) "No sir, she was not a fool in the field": gendered risks and sexual violence in immersed cross-cultural fieldwork. Profess Geogr 67(2):180–186
56. Patterson A, Kinloch V, Burkhard T, Randall R, Howard A (2016) Black feminist thought as methodology: Examining intergenerational lived experiences of Black Women. Departures Crit Qualit Res 5(3):55–76
57. Collins PH (2000) Black feminist thought: knowledge, consciousness, and the politics of empowerment, 2nd edn. Routledge, New York
58. MacCallum, D, Babb, C, Curtis, C (2019) Doing research in urban and regional planning: Lessons in practical methods, Routledge
59. Knapik M (2006) The qualitative research interview: participants' responsive participation in knowledge making. Int J Qualit Methods, 77–93
60. Braun V, Clarke V (2020) One size fits all? What counts as quality practice in (reflexive) thematic analysis? Qualit Res Psychol, https://doi.org/10.1080/14780887.2020.1769238

Observation for Data Collection in Urban Studies and Urban Analysis

Jason A. Byrne

Abstract Observation can help us better understand urban spaces, places and place-making. This chapter considers how observation can be used as a research method for gathering data to be used in urban analysis. Observation is more than just the act of looking. Observation requires careful and considered assessment of what is happening. The observer does not merely record information—they are also interpreting and analysing what they observe—albeit sometimes subconsciously. Observation can be divided into two main types—structured and naturalistic observation. We examine some of the key steps to be followed when collecting data using observation, referring to Australian and international case examples. There are advantages and disadvantages of observation compared to other methods and we consider some of these, including time, resources and data reliability. We also consider some important ethical issues related to various forms of observation, including deception and criminal activities. The chapter concludes with some thoughts about non-visual observation (e.g., soundscapes) and provides suggestions about how observation might evolve in the future as augmented reality, artificial intelligence and the 'internet of things' extend and expand the power of observation. Finally, some 'take home' messages are offered.

1 Introduction

Planners, urban managers and environmental officers—among other built environment professionals—regularly make decisions that affect the places where we live. Oftentimes, their recommendations, policy responses and decisions are made from a distance, sitting in front of a computer or in a committee meeting, and they are somewhat removed from the places affected by their actions. There is, however, no substitute for visiting a place to see how people interact with it, with each other, and with the environments found there. Indeed, observation has been called 'one of the most important methods of data collection' [1]. But undertaking observation is not as

J. A. Byrne (✉)
University of Tasmania, Hobart, TAS, Australia
e-mail: jason.byrne@utas.edu.au

© Springer Nature Singapore Pte Ltd. 2021
S. Baum (ed.), *Methods in Urban Analysis*, Cities Research Series,
https://doi.org/10.1007/978-981-16-1677-8_8

simple as walking into a social setting or a field site, switching on a recording device like a video camera or digital audio recorder, and making some notes. Key research design questions must first be answered, questions about where to observe, what to observe, how long to observe, how to record observations, and how to analyse the data. These sorts of questions are at the heart of the research. They are epistemological questions about what counts as valid knowledge—observation is not just about representing the world (what is seen) but also entails 'taking part in the world'. In other words, *how* we see affects what we observe [2].

For this reason, it is important to approach observation with rigour. How can we be sure that the things we observed during a site inspection are characteristic of the place we visited and how people experience that place and what they feel about it? What steps should we take to ensure that our observations are valid, and if required, could be defended in a court of law? How long does a place have to be observed—hours, days, weeks, months, or even years? Using what types of equipment? What protocols should be followed? Are there ethical guidelines for recording what is observed, reporting issues or protecting people's identities? Is using photography or video recording acceptable? And what if some of the behaviours witnessed are illegal? In this chapter, we will answer these and other important questions that should be considered by both the researcher and the practitioner when contemplating observation as a research method for data generation in urban analysis.

2   Observation as a Method of Data Collection

Observation has a long history as a research method. In the natural sciences, observation—as part of the scientific method, underpins a great many disciplines including chemistry, astronomy, ecology, physical geography and geology [3, 4]. In the social sciences, it has most commonly been employed in anthropology, sociology, psychology and geography [5]. In the social sciences, observation takes place in social settings that range from mundane and prosaic (e.g., counting traffic for a transport study) to the clandestine, esoteric and/or stigmatised (sex work or the activities of so-called outlaw gangs) [5, 6]. Observation evolved as a distinct method out of the work of the Chicago School of Sociology in the early to mid-twentieth century and continues to evolve as a research method [3].

Observation as a social research method is most commonly associated with participant observation, such as that conducted by anthropologists and sociologists in the study of cultural groups. Participant observation can be defined as "a nonquantitative method which claims to give privileged access to meanings through the researcher's empathetic sharing of experience in the worlds he or she studies" [7]. As a method used in ethnographic research, participant observation seeks to provide an 'insider' perspective as opposed to that of an 'outsider'. An insider is someone who is a member of an organisation or social group, observing from within, whereas an outsider is someone who is not a member of the organisation or group that they are observing [3, 5]. There has been a tendency for some to regard observation as easy, simple and natural—but as we shall show in this chapter, rigorous, reliable and ethical observation requires a careful research design [2]. Observation can be used as a stand-alone research method, or in combination with others (e.g., focus groups, interviews). Observation in some fields, such as urban ecology, is beginning to bridge the natural and social sciences, and increasingly is part of mixed-methods research (e.g., combining bird counts and park visitor surveys).

2.1 Types of Observation

Observation occurs along a spectrum from shallow or passive observation—such as counting traffic or birds, to deep or immersive observation—such as participating in a complex social scene and recording and interpreting what happens [8, 9]. Observation can thus be reductive and quantitative (seeking to enumerate) or inductive, experiential and richly descriptive (seeking to elucidate)—essentially these are different 'ways of seeing the world' [1]. An example of the former is Paul Knox's early work on environmental quality assessment whereas Clifford Geertz's study of Balinese cockfights exemplifies the latter [10, 11].

Some scholars also distinguish between controlled and uncontrolled observation—or structured and unstructured observation. Controlled observations are often used in experimental research designs—and tend to follow an explicit series of steps about what, where, when, who, how and how long to observe—as a way to isolate a variable that is being studied (e.g., park use). In contrast, uncontrolled, naturalistic observation takes place a bit like a vacuum cleaner—where as many details about what is being observed are recorded, and then a series of filters are applied to sort and categorise the observations for subsequent interpretation [9, 12]. A key purpose of unstructured observation is to identify norms, rules, explicit meanings and implicit meanings. Much of human culture consists of 'unspoken rules' that we learn in our ordinary lives (see Box 3). In unstructured observation, the researcher seeks to learn these implicit rules and to make them explicit.

2.1.1 Structured Observation

Structured observation, as the name implies, is undertaken in a carefully controlled fashion. It typically seeks to control for different variables such as time of day, day or week, season, place and the like. Typically, structured observation employs a schedule for recording activities. The schedule identifies the various activities to be recorded as well as how they are to be recorded [1]. This type of observation attempts to manage bias by using a method that is repeatable and verifiable, where the observations can be reliably used as an accurate record of what is being observed (see case example 1). In a case example on urban trail research presented later in this chapter, the structured observations of the trail and the trail users were carefully controlled. The data gathered in structured observation are often related to counting something—people, activities, encounters between people or people and objects, and therefore can readily be managed in a tabular form—for example, a Microsoft Excel spreadsheet (see Box 1).

Structured observation of park users

How people interact with urban spaces is a key concern of many built environment professions. Architects, urban designers, planners and transport engineers can benefit by better understanding human behaviour. Yet the provision of public facilities such as libraries, bicycle paths, swimming pools and parks remains comparatively understudied in evaluation research. Are people using these spaces as intended? What conflicts occur in their use? And what design solutions might resolve some of these issues? Many people will be familiar with the experience of walking past a public park and thinking that it seems empty and must not be utilised. But how can we truly know? Observation research can help us better understand park use.

In research on parks and squares in European cities, Barbara Goličnik Marušić and collaborators developed a rigorous way of undertaking structured observation of park users and how they interacted with park features [13, 14]. Troubled by a relative paucity of knowledge about how people use park facilities, they devised an approach to observation that combined observation and spatial analysis. The parks were divided into sections for observation. Observers spent 10-min intervals, across the day, observing users. The day was divided into four periods—10 am to 12 noon; 12 noon to 2 pm; 2 pm to 4 pm; and 4 pm to 7 pm, during the month of May (European summer). Information about the type of park user (male, female, older, younger), the activities they were undertaking (pushing a pram, cycling, skateboarding, dog-walking), and park features (trees, paths, fountains etc.), were recorded on paper maps of the parks, and then were transferred to a geographic information system (GIS) for analysis.

What the researchers found is that park features are indeed significantly associated with different user groups and activity types. People like to sit on walls and watch passers-by. People sitting on the grass usually do so at least 5 metres away from a path and also tend to spatially distance themselves from other users—especially groups, who can have 20 m buffers. Trees are attractive for people who like to sit on the grass. Some spaces are used intensively whereas others are relative dead zones. And activities vary across time of day and day of week.

But their research had some limitations. It did not account for seasonal variations. It also missed key activity periods in the early morning and at night. Researcher fatigue means that there was, inevitably, error in their observations and because people were only observed, not interviewed, their motivations and choices could only be inferred. However, some potential bias was managed by recording activities in the same month in two different years, and park areas were observed on different days, at different times and under different weather conditions.

2.1.2 Unstructured/Naturalistic Observation

Unstructured observation is a method suited to settings and situations where the researcher is trying to record as much as possible—guided by intuition, past experience or both. Rather than trying to eliminate subjectivity or manage bias, this form of observation is deeply subjective and accepts bias as somewhat inevitable. The researcher acknowledges that their own subject position—shaped by their upbringing and their ethno-racial and socioeconomic status (e.g., religion, gender, race, income level, sexual preference etc.)—will affect what they observe and how they observe. This type of observation is usually accompanied by detailed field notes and sketch maps, as well as elements of conversation that are either recorded verbatim at the time (as best as possible) or recalled later [1]. Field notes are essential for naturalistic observation and are used to record events, people, places, feelings, concerns, activities as well as the researcher's own personal reflections, mental notes and memory joggers [9].

Participant observation is typically a form of unstructured observation. The observer is a participant in the activity or social scene being observed. The purpose is to obtain a deep, insider perspective. Often the aim of participant observation is to corroborate (or refute) what participants say they do, versus what they actually do. Participant observers may, or may not, tell those who they are observing that they are being observed. Sometimes, it is not possible to be explicit about the act of observation (see discussion on ethics below), but if a researcher has sought to conceal their activities in covert observation, they risk negative sanctions or even violence if they are discovered (see Box 2). The participant–observer is, therefore, faced with the dilemma of how to record what they are observing. Having a conspicuous recording device or a notebook would potentially change people's behaviour, revealing to others that they are being observed. Typically, some form of abbreviated notetaking is required (e.g., dot points in a notepad on a mobile phone), and these notes are expanded as soon as an opportunity allows. The purpose of field notes is to record rich details and to describe what was observed without making value judgements [9].

Because participant observation seeks an insider's perspective, if the researcher is not already a member of the social setting they are researching, they will need to gain access (see Box 2). This is usually achieved with the assistance of one or more 'informants', who are critical brokers of knowledge and information. An informant is usually told by the researcher that they are undertaking research. Informants can provide additional knowledge by answering follow-up questions and offering personal insights [3, 8]. Sometimes informants may be asked or may volunteer, to take photographs or video record a setting, though there are ethical issues to consider, which we address later.

Data collected during participant observation are often complex—field notes, sketch maps, photographs, videos, audio-recordings, personal reflections and the like. For these reasons, there may be multiple forms of analysis required. For example, field notes could be analysed using software such as Leximancer or NVIVO to undertake a text analysis or content analysis, coding keywords and phrases and then calculating their frequency—and also assessing their context (see Chap. 13). Photographs might be analysed using content analysis techniques, identifying both the main elements in each photograph and key cross-cutting themes across different photographs.

Increasingly the distinction between structured and naturalistic observation is becoming blurred. For techniques such as participatory photo mapping, for example, structured and unstructured observations are melded. Participants take photographs of the places they observe and the researcher interprets these images according to what the images contain and what participants say about the images, as well as using spatial data about the images to build a more complete understanding of the phenomenon being observed, via content analysis and mapping (spatial analysis) [11]. When combined with the use of tracking technologies, such as global positioning systems (GPS) and cellular network data from smartphones, the researcher can analyse not just where the photos were taken, but where the participants went when choosing to take their photographs—and even their emotional responses to different settings if participants were equipped with skin sensors, though this can also raise privacy concerns (see ethics) [15].

Participant observation in an adult store

An important consideration in observation research—especially participant observation—is how to gain access to the social setting [3]. When observation is overt, and the researcher makes it clear what they are doing, people can change their behaviours. This can be quite difficult if the setting is considered to be clandestine, deviant or morally questionable—so sometimes researchers undertake covert observation. An example is adult stores.

While adult stores (sex shops) are commonplace in many cities, understanding how they function and the legitimate planning and environmental management considerations that need to be taken into account (parking, noise, signage, location, operating hours) versus moral judgements about their appropriateness, is a matter that could be better understood via observation research. Some activities will be legal (e.g., selling sex toys) whereas others may be illegal (e.g., prostitution). In Australia, there are examples of council employees who are contracted to monitor adult stores, massage parlours, brothels and other sex-related land uses to ensure that human trafficking is not occurring and that health and planning regulations are followed [16].

For the researcher, gaining access to a complex social setting such as the adult store requires careful research design. In observation work undertaken in an adult store in Washington DC, USA in the 1980s, the researcher gained access by securing employment as a night operator. This gave them a legitimate reason for being in the social setting, but also meant they had to negotiate complex issues related to illegal activities that occurred during their covert observation [16]. Because the observation was undertaken furtively, it also raised ethical concerns. However, the researcher was able to access reliable data that otherwise would have been off-limits. Evidencing this, they observed other researchers being told lies by sex workers because they were not trusted or had not built rapport.

2.2 Scale in Observation

Observation as a research method takes place across the full spectrum of scales—from the body to the globe. The techniques used in observation vary accordingly. While it was less common in the past, research is also beginning to employ multiple techniques that enable observation across different scales within the same study.

2.2.1 Observing from a Distance

Observation can occur from a distance—a long distance. Satellite observations have been around since the 1950s, but with the availability of comparatively inexpensive computing and satellite technology, access to satellite data has never been easier. High-resolution data from satellites such as Landsat and Sentinel give scholars and practitioners the opportunity to observe the Earth's features and human activities at a global scale, and across time periods up to decades. Good examples are the observations of polar sea ice and changes in vegetation cover that have provided compelling evidence of climate change. Satellite observations are increasing in resolution. Two decades ago, the level of resolution possible was around 30 m—nowadays it is possible to achieve accuracy within 15 cm. These high-resolution, remote sensing observation capabilities are especially important for planners, environmental managers and even emergency services personnel. For example, it is now possible to assess the health of urban infrastructure such as bridges using satellite technology for remote observation, and to rapidly locate damaged buildings following a flood or earthquake and even to identify the maximum cooling effect of different types of green infrastructure to combat urban heat islands [17, 18]. Unoccupied aerial vehicles (UAV) or drones also provide unprecedent opportunities for observing urban activities, from the scale of an individual yard to multiple suburbs. A good example is thermal imaging to detect heat island variations. However, this technology is not without issues—increasingly there are concerns over invasion of privacy and/or amenity impacts from noise.

2.2.2 Getting up Close and Personal

More typical in social research is the observation that occurs at the sale of the body, home, workplace and neighbourhood. Such observation might entail observing workers in a factory assembly line to learn about workplace practices, aged care residents in an older people's home to observe carer behaviours or students in a classroom, to observe learning behaviours. Some forms of observation can occur in very small spaces such as lifts and public toilets (see Box 3), though there is a range of ethical issues that are triggered by the observation of more intimate spaces, and strict protocols need to be implemented to ensure that legitimate observation does not become voyeurism [3] (Fig. 1).

Observing people's behaviour in lifts and public toilets
A key task of naturalistic observation is to make explicit the implicit rules of behaviour that shape our societies and social settings. Let us consider two very urban spaces—the lift (elevator) and the public toilet.

Fig. 1 Choosing where to stand at a public urinal is normally guided by unspoken rules

Nobody gets into a lift and turns to face the other occupants with their back to the doors. Nor do we sing, sit down, play music, eat food, use an electric shaver—or stare at others for extended periods of time. If the doors open and someone wants to get in, the lift occupants shuffle around to create more space, in a carefully orchestrated dance with limited physical contact. The lift is a liminal space of small gestures, orderly and predictable actions, hushed and brief encounters, social niceties, cursory pleasantries, limited eye contact and furtive observation [11]. It is also a space where people almost unconsciously know where to stand, so as not to block others ingress or egress but also to protect their own personal space. There are tacit rules of distance-keeping, to maintain equidistance. There are other rules too, based on avoiding conflict or unwanted attention or mistaken signals. Where is it OK to look, for how long and at what? When is it OK to talk, about what and for how long? Most city-dwellers could answer these questions readily. Who taught us these rules? And what happens if we break them?

In a study of men's public toilets, Elliott Oring made some of these unspoken rules explicit [19]. A series of norms guide people's behaviour around public toilets—from approaching the building or room, through to standing to urinate, washing hands and exiting. As with many other bodily functions, urination creates a semi-taboo space. Language used tends to employ absent referents—'going to the bathroom' rather than 'taking a piss', and spending long periods of time in a public toilet only happens in some specific social settings—for example, high school students 'wagging class'.

As with our lift example, when a man enters an unoccupied toilet to urinate, there are tacit rules about where to stand—in the corner, furthest from the door, protecting personal space and privacy. If there is someone already occupying that spot, then the opposite corner is selected. If both of these positions are taken, then a position between the two is chosen. Wherever possible, a minimum distance of one person is maintained. If personal space cannot be maintained, it is common to choose a stall instead. Clean urinals are preferred over dirty ones. The only time these rules are violated is when two men know each other, and may thus stand closer together. Talk between strangers is limited, but acquaintances may have a brief conversation, and conversations are louder in volume, never whispered. Eye contact is avoided and the person urinating will either look straight ahead or down towards their feet.

3 Steps in Observation

Many research method texts point out that unlike other methods, observation research does not follow a series of prescribed steps. Nonetheless, it is possible to identify some key steps that need to be followed, including choosing the setting for observation, gaining access to the setting, deciding what to observe, and choosing when and how to observe. Although we discuss analysing data and ethics as separate headings, obtaining ethics approval and data analysis and reporting can also form key steps in observation research.

3.1 Choosing the Setting

The choice of the setting for observation research does not happen by accident. The chosen setting will reflect the purpose of the research. If the goal of the research is to understand how residents use common space in an apartment complex, then the setting will necessarily be an apartment building. However, in this example, some choices still remain—is the apartment public housing or market-based? How many units should the building contain—6, 20, 100? and should the setting include a range of common spaces—corridors, garages, outdoor areas, a pool? Similarly, if the purpose of a study was to assess how women use parking garages to assess gendered use of urban space, then questions will need to be answered about whether it is above-ground or below-ground, public or private, lighting levels and the like—as well as location, such as a CBD or a suburban shopping centre.

Sometimes a setting will be unfamiliar to a researcher, and will require considerable background research, including multiple site visits and potentially even accessing archival documents or undertaking a review of policies and plans [2]. Some scholars contend that it may be better for a researcher to be unfamiliar with a social setting, as this enables a sort of detachment that facilitates ready observation, whereas familiarity with a social setting can mean that an observer slips into being a fully immersed participant and loses track of what they are observing [2].

3.2 Gaining Access

Some social settings are public spaces where no special permission is required for access. These include beaches, parks, streets and public pools—though parts of these sites may be off-limits (such as change rooms), and extended periods of observation may still require approval. Other settings are semi-public but access is regulated to protect their users (e.g., schools, hospitals, airports, churches, public housing), requiring permission to undertake research. Some social settings are heavily restricted (e.g., military bases) whereas others occupy a more liminal space between public and private (e.g., shopping centres). In the case of shopping centres, it is possible to undertake observation without permission, as both a shopper and an observer, but some activities such as taking photographs or taking videos would, strictly speaking, require the permission of centre management. This will likely also apply to airports and potentially train stations, depending on the nature and duration of observation—though taking a selfie may be one way to circumvent that requirement, and would not draw the attention of other users (see Fig. 2).

Other social settings may require the assistance of an informant to gain access. These include social or sporting clubs, street workers, an airport lounge, a religious group, street art sites and gangs, among others (see Box 2). Because observation may occur over extended time periods, it will be important to manage relationships with informants carefully, so as not to betray their trust. Informants will likely confide in the researcher after trust is earned, and although the researcher can walk away from a setting after the research is complete, the informant will likely continue to remain in that setting, and can suffer consequences such as ostracism, loss of income or even physical violence if they are seen to have given away secrets or restricted information. We return to this point in our discussion of ethics.

3.3 Deciding What to Observe

One of the key challenges in observation is what to observe. Some of the intended subjects of observation are obvious. For a chemist—it is chemical reaction; for a geologist, rock formations; and for an archaeologist, ancient ruins. But the subject of observation for a social researcher is not always so easy to discern. If a researcher is studying a surfing subculture, for example, what aspects of surfing should be observed? Should the researcher just focus on the line-up where people are catching waves, or are other places also important, such as the carpark, the beach, or even a café or pub frequented by surfers? The answer to this question is important as the

Fig. 2 In some settings, photography requires permission, though selfies may be an exception

decision of where to observe affects the quality of the data that is obtained. So if a researcher only focused on the line-up, they would miss a lot of rich interactions that occur on the beach and in the parking lot, and would also miss potentially critical information for their study [20, 21]. Observation requires detailed note-taking and at the beginning of observational research it is not uncommon to take copious notes, while figuring out what details are important and which ones are not. Sometimes small details and subtle behaviours are important. For this reason, researchers need to allow sufficient time for observation and may also need to consider discretion (see Box 3).

For structured observation the question of what to observe is a bit easier to answer. The research questions will dictate the subjects and objects of observation. But as noted earlier, a rigorous research design is essential to ensure that the researcher is clear about what is being observed and how it is to be recorded. Oftentimes, the time and expense of research mean that there will not be a second chance to follow-up on missed details, especially if the observation is undertaken in a remote or clandestine location (see case study 1).

3.4   Choosing When and How to Observe

The recording of observations differs between structured and naturalistic observation. For structured observation, the researcher will complete a schedule, filling out information such as date, time, weather conditions and other relevant information. Structured observation may take place continuously or using a sampling approach. In the case of the former, a researcher may set up a video camera and record a continuous period of time—hours, days, weeks etc., depending on the capacity of data storage and the research question(s) the researcher is seeking to answer. For sampling a setting over a specific period of time, the researcher will need to first familiarise themselves with a site and potential diurnal and seasonal variations in use. For instance, a public park will have a rhythm of use associated with time of day and user characteristics (see Box 1).

For example, joggers may be out in the morning before work and again in the evening after work. Parents of young children may visit during mid-morning and mid-afternoon, based on the sleep patterns of their child and the availability of companions. Older people may avoid the heat of the day. Dog-walkers will have a natural rhythm based on working hours and the needs of their companion animals. To undertake a sample of park users at 2 pm on a weekday for observation would potentially miss a range of different users and a range of different behaviours. Teenagers may only visit on the weekend, after school with peers or even at night, after everyone else has gone—especially if they are drinking or engaging in sexual activity. Similarly, if a park is a gay beat, and the purpose of the research is to understand patterns of use of homosexual visitors, then the researcher will need to be familiar with visitation patterns of park visitors who will likely not want to be observed—and to have a respectful and appropriate way of observing, such as accompanying park rangers during their shifts [22]. The researcher should thus familiarise themselves with a site first, visiting often, across different days, weeks, seasons etc., to establish the variations in use before undertaking observation. If this is not possible due to budget and resources, then a sampling protocol should be developed to allow as much coverage as possible in the time available—accounting for different times of the day, days of the week and if possible seasonal variations—see case study 1 [23].

For naturalistic observations, the recording of data is more complicated, as noted above. Aside from the challenges presented to a researcher by operating in a complex social setting (e.g., a pub, library, beach, child-care centre), the researcher will need to make decisions about what time periods they will observe. For a museum, for example, this may be straightforward—the researcher will seek to record activities during opening hours. But there will still be challenges about the time a research can be immersed in a social setting, due to issues such as fatigue, the need to eat, sleep and go to the toilet, and because there may be long periods of time when nothing much is happening (for instance a beach on a cold and wet winter day), and so a researcher will need to have effective ways to manage boredom to ensure important information is not missed due to a lapse in attention.

4 Analysing Observation Data and Presenting Results

Given the topic of this book is urban analysis, it would be remiss not to discuss how data collected using observation is analysed. Again, this will depend on whether the data is derived from structured or naturalistic observation. The key to data analysis is the search for patterns. Researchers seek to identify similarities and differences in their data, as well as clustering patterns—in space and time and then report the significance of their findings. If the observation data included sketch maps, or used tracking technology, then data can also be analysed using spatial analysis, such as using a geographic information system (GIS) [24]. We address these types of analysis in turn.

4.1 Analysis of Data from Structured Observation

Most structured observation is about quantifying a phenomenon—in other words, it involves counting and summing up results. If we return to the earlier example of park use (see Box 1), the different types of users, and the locations where their activities were observed are tabulated and summed. Statements can then be made about the clustering of certain activities—under trees or along walls. If the sample is large enough, and data have been collected in a systematic way, it may be possible to undertake statistical tests—looking for significant differences by time of day, day of week or season, and then to test for differences between different user types (e.g., walkers, joggers, cyclists) and potentially variations by socio-demographic characteristics (e.g., age, sex, group size).

The purpose of these tests is to understand if the data collected may be generalisable for a broader population, and that findings are not just the result of random chance. At a minimum, analysis of such data will require calculating descriptive statistics—frequencies, mean, median, range and standard deviation [25]. But if the data allow, and the analyst is confident in undertaking statistical analyses, inferential statistics can be used to test the generalisability of findings to the broader population. Statistical tests include t-tests (to compare averages), analysis of variance (ANOVA), and chi-square tests (among others), using software such as R or SPSS. Findings are typically presented using tables and charts.

4.2 Analysis of Data from Naturalistic Observation

In contrast to structured observation, the analysis of data collected through naturalistic observation will be more descriptive. The researcher will still be looking for patterns in their data, and even for variations by age, gender, income and the like, but will not be able to generalise these to the broader population.

The analysis of unstructured observational data is usually done using coding and text analysis. Codes are descriptive labels that are attached to 'chunks of information' that might be individual words, sentences or even paragraphs [25]. But codes can also be applied to images—classifying them according to their contents (e.g., trees, people, buildings) [26]. What matters most in naturalistic observation are not the things that were observed, but their *meaning*.

5 Ethical Issues in Observation

Both structured and naturalistic observations require the researcher to carefully manage some potentially complicated ethical issues. Privacy, confidentiality and anonymity are important considerations, so too are power relations, safety and well-being. Early observation research such as that undertaken by Cressey in his study of Taxi Dance Halls involved deception and covert observation. Cressey would often assume a pseudonym, invent an identity to get closer to research subjects to elicit information that may otherwise not have been disclosed, and undertook observation without others knowing they were being studied [27]. Nowadays such activities are likely to be regarded as unethical. Researchers who wish to undertake observation research will usually require the written approval of their institution's Human Research Ethics Committee (HREC). There are, however, some types of observation that will likely not require an ethics approval, such as undertaking traffic counts. If the observation also involves animals, then approval may also be required from the researcher's Animal Ethics Committee.

In Australia, researchers undertaking social and medical research are bound by the *Australian Code for the Responsible Conduct of Research* [28]. For example, taking photographs of people for research purposes, where their faces are identifiable, will oftentimes require their written informed consent—or the subsequent masking of people's identities (see Fig. 2) [2]. An informant's identity will need to be protected to ensure they are not harmed—physically, financially, emotionally or psychologically. When notes are made about people in a social setting, their identity should be protected too, and if interviews are conducted, pseudonyms should be assigned unless an informed consent document explicitly states that a participant freely and openly gives their permission for their identity to be made public. If the observational research involves children, people with a disability, Aboriginal people and other potentially vulnerable groups, there will be heightened attention to avoiding potential harm.

Some observational research will require the researcher to make difficult decisions. There may be cases where covert observation is justifiable, but it will need to be carefully weighed against the need for participants to willing consent to participate, knowing the potential benefits, harms, risks and implications of participating [27]. What is the obligation of the researcher if observing illegal activities such as tagging, drug use, assaults, racial vilification and the like? Many scholars contend that as a citizen, the researcher is duty-bound to report illegal activities to the authorities [2]. But if the observation is covert, and researchers did not obtain informed consent, it could be argued that the people engaging in nefarious activities would not have done so if they had known they were being observed. There are many examples of covert observation research where we have learned a great deal about human psychology, the ways that power can be abused, and about dysfunctional organisations and institutions—with substantial social benefits. But covert observation basically robs participants of their right to voluntarily consent and is thus a violation of the ethical principles of transparency, trust, disclosure and respect. Ultimately such matters need to be discussed with the HREC prior to commencing research.

Reciprocity and power dynamics are other important ethical considerations. If the researcher establishes relationships with other people as part of the research, these should be maintained as far as possible after the research concludes. The researcher should also make the research findings available to the people who have participated in the research. In many cases, participant observation entails the researcher making judgements and interpreting what they see, and some would argue that this comes with the obligation of letting informants and participants 'tell their side of the story', to at least elaborate or offer points of clarification. In some cases, the knowledge shared with a researcher may be culturally important, so the researcher will need to consider how the research findings are reported in ways that are culturally safe and respectful [2].

6 Observation in Practice

Discussions about research methods can oftentimes be very abstract. For this reason, the previous discussions about the types of observation, steps in observation and ways of analysing data are illustrated by two case study examples—the first is research that the author was involved in while a research assistant in the USA and more recently, research the author collaborated in as a professor in Australia.

Case study 1—Research on urban trail environments

Are residents who live near an urban trail more likely to be physically active? A study by Jennifer Wolch, Kim Reynolds, Michael Jerrett, Donna Spruijt-Metz and others in the early 2000s was one of the first to try to comprehensively answer this question, by assessing the interactions between characteristics of the built environment and residents' level of physical activity [29]. At the time, there was a growing realisation that the Western world was in the midst of an obesity epidemic. The built environment was suspected of playing a strong role in either constraining or enabling physical activity. Researchers speculated that automobile dependence, low-density neighbourhoods, segregated land uses and inadequate access to recreational facilities such as parks and green spaces were contributing to sedentary lifestyles [30]. The research team included academics from geography, planning, public health, psychology, epidemiology and public policy, who were seeking to extend our understanding of the correlates of built environments and physical activity. One of the challenges facing the research team was how to separate out the effects of people's behaviours, features in the built environment, land use planning policy and trail governance and management. They turned to observation to provide answers.

The research team devised a mixed-methods approach that consisted of two types of observational activities, in combination with spatial analysis using a geographic information system (GIS), a telephone survey of residents from diverse social backgrounds from trail-adjacent neighbourhoods, and a sample of people fitted with accelerometers to record their physical activity levels. As shown in Fig. 3, two types of observations were recorded. First, the researchers modified a built environment auditing tool used by Billie Giles Corti and colleagues to assess the physical characteristics of walking and cycling facilities in Perth, Western Australia [31]. This observational instrument was adjusted for auditing the physical characteristics of multiple-use urban trails (e.g., walking, cycling, roller-blading etc.) in Chicago, Dallas, and Los Angeles USA. Two members of the research team walked the entire length of the trails (which were over 20 km long), making detailed observations and recording information they observed about adjoining land uses, trail condition, trail construction materials, signage, trail-side facilities (e.g., benches, fountains, toilets), vegetation type and density, entry and exit points and the like). This was undertaken at 800-metre intervals along the entire trail—for all three trails—using audit sheets. The two auditors' assessments were then compared using a kappa statistic to evaluate their inter-rater reliability [3].

Fig. 3 Observation of urban trail use in Chicago, Dallas and Los Angeles, USA

A separate team of researchers rotated in shifts, dispersing to different segments along the trails, observing and recording what people were doing on the trails (e.g., walking, running, bird-watching, walking a companion animal, horse-riding). These trail-user observations were important, as they potentially allowed the research team to compare how the physical characteristics of the trails and adjoining land uses affected how people were using (or avoiding) the trails. The trail user observations also collected basic socio-demographic information about the users—were they male or female; young, middle-aged or older? And the observers also recorded the day of the week, time of day, trail segment and start and finish times. Recording duration was for 15 min—to manage observer fatigue.

Some key issues emerged during and after the study. While it would have been very useful to know the ethno-racial composition of trail users, this could not be determined through observation. Making a judgement about whether someone is White or Latino is not an appropriate way of evaluating their ethno-racial characteristics, for two reasons. First, race and ethnicity can be used as the basis for discrimination; it is important to know how someone self-identifies, rather than ascribing to a category based on the perceptions of the observer. Similar issues occurred when evaluating the age of trail users—if a user appeared to be at the edges of a category such as middle-aged vs. older, was the judgement of the observer sufficiently accurate?

And accurately recording all the trail users proved to be challenging on busy segments of the trail, or when observers were tired or distracted. The researchers considered using an automatic counting device, but this would not have provided information about the type of trail users. Observing trail characteristics was also beset with some challenges. While the presence or absence of features such as rubbish bins, signs or fountains was relatively straightforward, accurately recording characteristics such as 'somewhat littered versus very littered', 'poorly maintained versus very poorly maintained', or 'attractive versus very attractive' created some situations where there was a low level of agreement between the trail auditors. This is because the evaluation was subjective and a matter of a value judgement. Some would suggest that clearer categories or more rigorous training of auditors would produce more accurate results, but this is a dilemma of observation—is the accuracy reflecting level of agreement between auditors and the instrument, or between the observer and the so-called real world?

Case study 2—A new community garden for students and residents

In our second case study example, the University of Tasmania Facilities Management Division was considering developing a community garden in a student accommodation complex (see Fig. 4). The University had been criticised because the common areas were perceived to be cold, lifeless and alienating. A community garden was imagined as a good way to bring students into closer contact with each other and the surrounding businesses and residents. But there are many examples in the literature of where community gardens have failed because they were poorly designed and/or did not have secure tenure and a commitment for ongoing management [32]. The University wanted to avoid this, so it commissioned research to better understand how the space was used prior to developing the garden.

The research team devised a mixed-methods approach to obtaining data about how the common spaces in the building were used, and student attitudes towards the University. This included a survey of students in the building, interviews with students and surrounding businesses and residents, focus groups and visual structured

Fig. 4 Observation of community garden activation project in university accommodation. *Source* The watering can image is courtesy of Bodhi Diaz-Icasuriaga, University of Tasmania. All other images were taken by the author

observations. The goal was to identify and understand the barriers and enablers to the successful establishment of a community garden.

The non-participant observation had two elements. A member of the research team followed a pre-determined circuit through the site, at different times of the day and on different days of the week, using a schedule to code the different types of people using the site and the activities they were undertaking. This was augmented by sampled recording from two fixed video cameras. Site users were notified that recording was in progress via well-placed signage. The public life observation consisted of 7 h of observation in total, at various time periods across four different days. The video footage totalled 14 h, across various time periods, and 4 days. Members of the public utilising the spaces included student residents, nearby residents of the broader community and business owners as well as passers-by.

The observational data and video footage were analysed using interpretative processes. This entailed viewing the video footage, coding its content (movement patterns and social interactions), and discussion by two of the research team. The main findings were that the space was comparatively underutilised at the time of the study. Aside from a group of parkour enthusiasts using the concrete structures, most people were either waiting for children, talking on the telephone, sitting, smoking, eating food or walking through the site. The findings informed the subsequent design and installation of planter beds and the establishment of a community garden (see Fig. 4).

7 Potential Future Trends and Directions in Observation Research

Advances in technology are changing what we observe, and how we observe in urban analysis research. As noted earlier in the chapter, the traditional approaches to observation such as watching and counting are being augmented by technologies such as GPS, GIS, skin sensors, remote sensing, and closed-circuit television (CCTV) among others. Rapid advances in computer processing speed and power mean that it is now possible to combine different types of sensing and observing in ways that were previously unimaginable. For example, combining location data from mobile telephones with CCTV images, participant interviews, site audits, participatory photo mapping and object-based recognition spatial analysis to identify trees and vegetation could generate very detailed understandings of urban park use [33]. In a similar fashion, crime data could be combined with built environment audits, satellite data, CCTV and skin sensors to build a comprehensive assessment of the walkability and safety of different built environments, segmented by population cohort characteristics—such as age, gender (dis)ability etc. [34, 35].

7.1 Non-visual Observation

While much of this chapter has focused on the visual aspects of observation, there are other forms of observation that are just as important in urban research and urban analysis. These include sonic (auditory), olfactory (smell) and tactile (touch) observations. Although vision and seeing tend to be privileged in urban geography and planning, built environments are rich in sound, smell touch and other sensory experiences [2]. Planners and urban managers often need to understand these aspects of towns and cities, and observation as a method has a role to play. Noise is a good example. An unexpected indicator of the extent of background urban noise that affects many built environments occurred during the lockdowns associated with the COVID-19 epidemic. As many cities went into lockdown, residents and urban researchers were struck by the simple sounds they could hear, sounds that were normally drowned out by urban activities and motor vehicle traffic. These 'background sounds' included bird song, people talking, children playing, and even the clank of pots and pans in kitchens at mealtimes [36].

Airports, busy arterial roads and industrial areas can generate substantial noise levels that can affect the health and wellbeing of residents and wildlife [37]. Some noise-generating land uses are managed through provisions in town planning schemes, for example, the requirement of a buffer from surrounding residential areas [38]. Others, such as airport noise are managed via flight path restrictions and operational curfews. But some sources of noise are less well regulated, including road traffic noise, people having parties with loud music and even air-conditioning, lawnmower and power-tool noise. These can disrupt people's sleep and affect residents' ability to concentrate. Prolonged exposure is associated with hypertension, anxiety and depression [37]. For these reasons, environmental managers will often deploy sound sensing equipment to observe, measure and monitor noise levels. But residents' observations also matter [39] and how to include them in observational research is an important consideration.

8 Conclusions

In this chapter, we have considered observation as a research method. We have noted that there are two main forms of observation—structured and naturalistic observation. They each have different approaches to data collection and analysis. Structured observation enables researchers to better understand what is happening at the places they observe. It is well suited to projects that have budget and time constraints, more limited human resources and which seek to understand how people interact with a particular built environment. Naturalistic observation is better suited to projects that seek to understand the meanings that people attribute to places, but is more resource-intensive, both in time and in the considerable effort required to gain access to a social setting and win the trust of informants. Both types of observation have ethical issues which we have discussed, and researchers need to consider these as part of their research design.

Key Points

- Observation is a well-used and important part of the urban analysis but it requires careful planning and design and a good understanding of the relative merits of structured versus unstructured observation.
- There are important ethical issues to consider, especially with regard to informed consent, privacy, power dynamics and leaving the social setting at the conclusion of the research.
- Technology is rapidly advancing our capacity to observe, in real time, across scales from the body to the planet. This presents some exciting opportunities for future observational research to move beyond the visual to the auditory and even olfactory, and will likely provide new insights into how people interact with built environments.
- As Savas has previously noted, the point of urban analysis is not just to observe, record, document and catalogue—but it also should effect positive change in the built environment [35]. For example, if a study documents a safety problem in a multi-story carpark, there should also be an effort to fix that problem and to learn from identified mistakes.
- And if research identifies profound ethno-racial, gender or age-based disparities, researchers should try to devise policy recommendations to redress these inequalities.

Further information

For those wanting to get further information about observation methods. These include:

- Ciesielska, M., Boström, K.W. and Öhlander, M., 2018. Observation methods. In Qualitative Methodologies in Organization Studies (pp. 33–52). Palgrave Macmillan, Cham.
- Clark, A., Holland, C., Katz, J. and Peace, S., 2009. Learning to see: lessons from a participatory observation research project in public spaces. International Journal of Social Research Methodology, 12(4), pp. 345–360.
- Evensen, K.H., Nordh, H. and Skaar, M., 2017. Everyday use of urban cemeteries: A Norwegian case study. Landscape and Urban Planning, 159, pp. 76–84.
- Jackson, P., 1983. Principles and problems of participant observation. Geografiska Annaler: Series B, Human Geography, 65(1), pp. 39–46.
- Odgers, C.L., Caspi, A., Bates, C.J., Sampson, R.J. and Moffitt, T.E., 2012. Systematic social observation of children's neighborhoods using Google Street View: a reliable and cost-effective method. Journal of Child Psychology and Psychiatry, 53(10), pp. 1009–1017.

- Park, K. and Ewing, R., 2017. The usability of unmanned aerial vehicles (UAVs) for measuring park-based physical activity. Landscape and Urban Planning, 167, pp. 157–164.
- Sharifi, E. and Boland, J., 2020. Passive activity observation (PAO) method to estimate outdoor thermal adaptation in public space: Case studies in Australian cities. International Journal of Biometeorology, 64(2), pp. 231–242.

References

1. Jones L, Somekh B (2009) Observation. In: Somekh B, Lewin C (ed). Research methods in the social sciences. Thousand Oaks, Ca.: Sage, pp 138–45
2. Kearns RA (2010) Seeing with clarity: undertaking observational research. In: Hay I (ed) Qualitative research methods in human geography, 3rd edn. Oxford University Press, Oxford, pp 241–258
3. Bryman A (2016) Social research methods, 5th edn. Oxford University Press, Oxford
4. Turkington A (2010) Making observations and measurements in the field. In: Clifford N, French S, Valentine G (eds) Key Methods in geography. 2nd. ed. Thousand Oaks, Ca, Sage
5. Johnson JC, Avenarius C, Weatherford J (2006) The active participant-observer: applying social role analysis to participant observation. Field Methods 18(2):111–134
6. Przemieniecki CJ, Compitello S, Lindquist JD (2020) Juggalos-Whoop! Whoop! A family or a gang? A participant-observation study on an FBI defined 'hybrid' gang. Deviant Behav 41(8):977–990
7. Platt J (1983) The development of the "participant observation" method in sociology: Origin myth and history. J Hist Behav Sci 19(4):379–393
8. Jackson P (1983) Principles and problems of participant observation. Geografiska Annaler: Ser B, Human Geogr 65(1):39–46
9. Glesne C (1999) Being there: developing understanding through participant observation. In: Glesne C (ed) Becoming qualitative researchers: an introduction. Longman, New York, pp 43–66
10. Knox PL (1976) Fieldwork in urban geography: assessing environmental quality. Scott Geogr Mag 92(2):101–107
11. Dennis SF Jr, Gaulocher S, Carpiano RM, Brown D (2009) Participatory photo mapping (PPM): exploring an integrated method for health and place research with young people. Health Place 15(2):466–473
12. Spradley JP, McCurdy DW (1972) The cultural experience: ethnography in complex society. Science Research Associates, Chicago
13. Goličnik B, Thompson CW (2010) Emerging relationships between design and use of urban park spaces. Landscape Urban Plann 94(1):38–53
14. Marušić BG (2011) Analysis of patterns of spatial occupancy in urban open space using behaviour maps and GIS. Urban Design Int 16(1):36–50
15. Shoval N, Schvimer Y, Tamir M (2018) Tracking technologies and urban analysis: adding the emotional dimension. Cities 72:34–42
16. Crofts P (2013). Not in my backyard: who wants a brothel as a neighbour? The Conversation: The Conversation, Available from: https://theconversation.com/not-in-my-backyard-who-wants-a-brothel-as-a-neighbour-21034

17. Jiménez-Jiménez SI, Ojeda-Bustamante W, Ontiveros-Capurata RE, Marcial-Pablo MdJ (2020) Rapid urban flood damage assessment using high resolution remote sensing data and an object-based approach. Geomatics Natural Hazards Risk 11(1):906–27
18. Yang G, Yu Z, Jørgensen G, Vejre H (2020) How can urban blue-green space be planned for climate adaption in high-latitude cities? a seasonal perspective. Sustain Cities Soc 53:
19. Oring E (1979) From uretics to uremics: a contribution toward the ethnography of peeing. In: Klein N (ed) Culture, cures and contagion: reading for medical social science. Novato, CA.: Chandler and Sharp, pp 15–21
20. Langseth T (2012) Liquid ice surfers—the construction of surfer identities in Norway. J AdvEduc Outdoor Learn 12(1):3–23
21. Fiske J, Hodge B, Turner G (1987) North Sydney: Allen and Unwin
22. Ablitt J (2020) Walking in on people in parks: demonstrating the orderliness of interactional discomfort in urban territorial negotiations. Emotion Space Soc 34:
23. Veal AJ (2011) Research methods for leisure and tourism: a practical guide, 4th edn. Pearson, Edinburgh Gate, Harlow
24. Páez A, Scott DM (2004) Spatial statistics for urban analysis: a review of techniques with examples. GeoJournal 61(1):53–67
25. Bell J (2010) Doing your research project. Open University Press, Berkshire
26. Prosser J (ed) (2000) Image-based research: a sourcebook for qualitative researchers. Routledge, Abingdon, Oxon
27. Cressey PG (1932) The taxi-dance hall: a sociological study in commercialized recreation and city life. University of Chicago Press, Chicago
28. National Health and Medical Research Council, Australian Research Council, Universities Australia (2018) Australian Code for the Responsible Conduct of Research. Canberra: Commonwealth of Australia
29. Wolch JR, Tatalovich Z, Spruijt-Metz D, Byrne J, Jerrett M, Chou C-P et al (2010) Proximity and perceived safety as determinants of urban trail use: findings from a three-city study. Environ Plann A 42(1):57–79
30. Sallis JF, Johnson MF, Calfas KJ, Caparosa S, Nichols JF (1997) Assessing perceived physical environmental variables that may influence physical activity. Res Q Exerc Sport 68(4):345–351
31. Pikora TJ, Giles-Corti B, Knuiman MW, Bull FC, Jamrozik K, Donovan RJ (2006) Neighborhood environmental factors correlated with walking near home: using SPACES. Med Sci Sports Exerc 38(4):708–714
32. Guitart D, Pickering C, Byrne J (2012) Past results and future directions in urban community gardens research. Urban Forestry Urban Greening 11(4):364–373
33. Ratti C, Frenchman D, Pulselli RM, Williams S (2006) Mobile landscapes: using location data from cell phones for urban analysis. Environ Plan 33(5):727–748
34. Bereitschaft B. Equity in microscale urban design and walkability: A photographic survey of six Pittsburgh streetscapes. Sustainability. 2017;9(7):1233 (online)
35. Ye Y, Li D, Liu X (2018) How block density and typology affect urban vitality: an exploratory analysis in Shenzhen China. Urban Geogr 39(4):631–652
36. Rogers D, Herbert M, Whitzman C, McCann E, Maginn PJ, Watts B et al (2020) The city under COVID-19: podcasting as digital methodology. Tijdschrift voor Economische en Sociale Geografie 111(3):434–450
37. Basu B, Murphy E, Molter A, Basu AS, Sannigrahi S, Belmonte M et al (2020) Investigating changes in noise pollution due to the COVID-19 lockdown: the case of Dublin, Ireland. Sustainable Cities and Society. 2020; (in press)
38. Offenhuber D, Auinger S, Seitinger S, Muijs R (2020) Los Angeles noise array—Planning and design lessons from a noise sensing network. Environ Plann B: Urban Analyt City Sci 47(4):609–625
39. Lagonigro R, Martori JC, Apparicio P (2018) Environmental noise inequity in the city of Barcelona. Transp Res Part D: Transp Environ 63:309–319

Discourse Analysis

Keith Jacobs

Abstract This chapter introduces urban and planning students to the methodological approach known as 'discourse analysis' that aims to uncover and provide insights on the way that language is deployed to maintain and exercise power. The chapter begins by explaining the origins of discourse analysis and the two most influential variants: 'critical discourse analysis' and analysis inspired by the writings of Foucault. The next section of the chapter considers recent examples of Australian scholarship in the fields of planning and housing where researchers have undertaken a discourse analysis. The main section of the chapter provides the reader with two examples of texts that consider the impacts of Covid-19 and the policy responses required. It is, in this section, that the reader can learn how critical discourse and Foucauldian approaches can be undertaken. The next section considers the criticisms that have been directed at discourse analysis and how these might be overcome. Finally, the chapter ends with six points as a way to summarise the main arguments that have been put forward.

1 Introduction

As the chapters in this book testify, urban and planning issues can be examined in a variety of ways. In recent years, researchers have often deployed an approach known as discourse analysis that emphasises the significance of language (written text, visual images and spoken) in the realm of social, political and economic life. This chapter explains the ideas informing the approach; discusses scholarship in the field; provides two worked examples of texts that you can use to extend your skills; offers advice on how to address criticisms that are directed at those who deploy a discourse analysis and finally identifies key texts for further reading.

While there are different ways that discourse analysis can be pursued, the aim is to provide insights into how language is deployed to advance political objectives. The fact that all urban locales are shaped by the social interactions that take place within

K. Jacobs (✉)
University of Tasmania, Hobart, TAS, Australia
e-mail: Keith.Jacobs@utas.edu.au

© Springer Nature Singapore Pte Ltd. 2021
S. Baum (ed.), *Methods in Urban Analysis*, Cities Research Series,
https://doi.org/10.1007/978-981-16-1677-8_9

them and by wider political and economic structures is the main reason why discourse analysis has been taken up as a method by urban and planning researchers. Consider, for example, the tensions that arise in city locations when new developments are proposed. Very often, there will be conflicts between different interest groups such as local residents and property developers about the scale of new developments and whether they infringe the amenities of existing residents. Planning authorities play an important role in mediating these conflicts and are required to make decisions as to what can and cannot be built. The conflicts between different interest groups are played out in media campaigns, public meetings and within the planning committees of municipal councils. Researchers are interested in these conflicts since they determine the spatial outcomes of the built environment and shape social and economic interactions. Discourse analysis is also used to shed light on urban politics more generally. An example are studies that consider how hegemonic ideologies that pathologise poverty can have a profound impact on how people on low incomes see themselves. We can be certain that discourse analysis will be used to consider the role of government in imposing restrictions on social interaction following the outbreak of COVID-19 as well as the many political decisions that have been taken to protect people's incomes when they have lost their jobs.

Within urban and planning studies those who deploy a discourse analysis are often interested in the ways that property developers and financial institutions such as banks are able to exert influence when decisions are made about the built environment. Usually, the discourse analysis pays careful attention to the production of written text, speech and visual images, etc. but there are also some investigations as to how texts are consumed (i.e. read, listened to and acted upon).

The methodological assumption that underpins discourse analysis is that the material, relational and spatial inequalities that are features of urban life are actualised through individual experiences. The ways that these inequalities are represented (in the form of texts, visual images and speech) is of particular interest to those who undertake a discourse analysis.

1.1 The Origins of Discourse Analysis

Discourse analysis is deployed in social science and humanity disciplines such as linguistics, media studies, political sciences, cultural geography, sociology and psychology. However, it was not until the early 1990s that its methods were taken up by urban and planning researchers. Up to that point, the methods of discourse analysis were associated with the writings of specialists working in the fields of linguistics and literary criticism. The interest in discourse analysis as a method of analysis can be traced back to the linguistic turn in analytical philosophy, in particular, the work undertaken in the mid-twentieth century by the philosopher Ludwig Wittgenstein. Amongst his claims was that the common assumption that language is simply a medium for communication and representation is mistaken. For Wittgenstein, language is performative and constitutes a social practice. His understanding of language use has radical implications for how we understand the way reality is apprehended. Meaning is not intrinsic to any word itself but can only be arrived at in relation to its wider social context and interaction. Wittgenstein's philosophy led to an interest in the ways that language operates and how texts and images have an ideological component. It is now recognised by all social scientists that language cannot be seen simply as uncontested medium in which ideas are transmitted. Rather language use is shaped by wider societal configurations that enable communication to take place.

And yet it is not just the influence of linguistically minded philosophers such as Wittgenstein that has encouraged the interest in discourse analysis in the field of urban studies. As important has been the burgeoning of new communication technologies such as the internet and mobile phones that have transformed not only the way people live their lives but also the conduct of governance. These transformations have continued at a fast rate, now that innovations in technologies such as machine learning are being commercialised. Consider, for example, the widespread use of communication platforms such as Zoom and Skype during the period of enforced lockdowns that followed the spread of Covid-19. It is also the case that computer software technologies such as *AntConc, NVivo, Lexus-Nexis* make it easier for researchers to trawl through digitally stored data such as newspapers, Twitter, Facebook and other media sources and report on their analysis.

Researchers who adopt a discourse analysis generally take a more critical understanding not only of the conduct of government and their practices but also the way new technologies are being deployed. The key research task for those who deploy discourse analysis is to make explicit the power relations and ideologies that shape contemporary societal and governmental practices. It is a claim of discourse analysis that in foregrounding power relations, new insights can be achieved on the way politics is conducted and its broader effects.

2 Approaches to Discourse

The two approaches that are most often adopted by researchers who choose to undertake a discourse analysis are: critical discourse analysis and Foucauldian inspired discourse analysis. Each approach is discussed in turn.

2.1 Critical Discourse Analysis

The most common approach has been developed by Norman Fairclough [1–4], Wodak and Mayer [5], and Van Dijk [6] and is known as critical discourse analysis (CDA). Fairclough is the most well-known of these authors, he claimed that

CDA provides a way to reveal the ideological influences and strategies that powerful agencies and actors use to advance their power. In his major study Discourse and Social Change [1], Fairclough proposed a framework for research that could encompass the micro-practices that are a feature of all human interactions, a middle-range or meso-level analysis that paid particular heed to the production of discourse and the way that it is enacted and also a broader political economy or macro analysis on the role of capitalism and the conflicts that arise from exploitation in the labour market. Fairclough proposed three strands of research activity are necessary for a critical discourse analysis. The first he called 'textual analysis', which entails an examination of the words and grammar that are a feature of all texts. The second strand of activity Fairclough defined as 'discursive practice', which entails an interrogation of some of the techniques that those who produce media articles, policy reports and speeches use to convince readers or listeners of their argument. Fairclough gave examples of exaggeration, irony and rhetoric. Finally, it was incumbent on all those who use critical discourse analysis to situate their investigation in a broader 'political context' (power relations and structures). See Fig. 1.

It is evident that much of the impetus for CDA stems from a concern that within the corpus of urban and planning studies, there is too little attention to the performative role of language. As Fairclough et al. [7] explain 'people not only act and organise in particular ways, they also represent their ways of acting and organising and producing projections of new or alternative ways, in particular discourses'. In making this observation, Fairclough establishes actions, production and representation as to the field for investigation. When it comes to examining the production of documents and other texts, CDA is interested in the way that authors deploy rhetorical strategies. He pays attention to the use for example of metaphors, tropes, stereotypes, hyperbole and fallacies to buttress arguments (what Fairclough referred to as intertextuality).

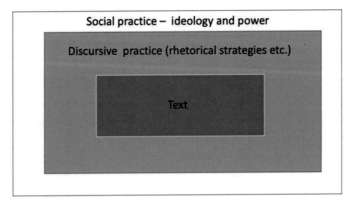

Fig. 1 Conceptual framework for discourse analysis (Adapted from Fairclough [1])

2.2 Foucauldian Inspired Discourse Analysis

The second approach that is used by urban researchers is known as Foucauldian informed discourse analysis. The approach is based on arguments espoused by the French historian of ideas Michel Foucault [8–11]. In his works, Foucault set out to explore how competing discourses influence and shape societal relations. Much of his study attended to the role of government and the way that governments enact discourses to secure their legitimacy with the public. For Foucault, this required a study of power relationships and he used the term 'regimes of truth' to describe the ways in which particular understandings are accepted. Through his historical analysis, Foucault noted that at certain periods there were considerable shifts in societal power relations due to changes in the economy and as a consequence, the relationship between government and the population were reshaped (see Dean [12] and O'Farrell [13]). Foucault used the term 'governmentality' to denote the practices and techniques deployed by the state to exercise power. Foucault's arguments about the performative aspects of discourse are set out in *Madness and Civilisation* [8]; *The Order of Things* [9] and *Discipline and Punish* [10], In was here that he discussed how concepts of madness have served to extend the reach of medical intervention; how the academic disciplines such as linguistics, biology and economics have evolved; and the way that surveillance tactics established in prisons extend to other areas of government.

It is important to understand that discourse analysis inspired by Foucault does not aim to uncover surface meanings as a first step to probe some deeper underlying meanings. Foucault's project is primarily to reveal the power relations that are exhibited in the text and the ways that discourses are actively produced (see Rose [14]). Foucault provided no instructions on the methods of discourse analysis although some approaches are explored in his publications (see Foucault [11]). The lack of guidance has prompted some authors to offer a framework. One example is Anderson [15], he identifies four strategies that are implicit in Foucauldian approaches. Anderson labels these as 'archaeological discourse analysis' which aims to identify the way that different truth regimes become hegemonic. 'Genealogy' pays attention to different historical periods and how discourses become replaced or superseded. The third strategy Anderson calls 'self-technology' that attends to the way that the practices of discourses are ordered. Finally, the strategy that Anderson terms 'dispositive analysis' explores the logics and rationalities used by different institutions.

3 Australian Urban, Planning and Housing Scholarship

In the field of Australian urban and planning studies, it is recognised that access to and distribution of resources are uneven and largely determined by wealth and social status. As the Australian government put in place infrastructure measures to ameliorate the challenges that arise from population increase, particular attention has been paid to the way that terms such as 'community', 'neighbourhood', 'liveability' are deployed to promote a vision of city life. Scholars are interested in charting the reasons why these terms are deployed and identifying whose interests they serve. As mentioned, a feature of discourse analysis is to establish the way that the meanings of these terms have shifted over time, how they are contested and why they stay unsettled. One clear example is the term 'planning' in which Australia is subject to different interpretations. On the one hand, there are those who see planning authorities as an impediment to commercial development and lobby to reduce the influence of planning in the decision-making processes. On the other are those seeking to extend the capacity of planning authorities to protect neighbourhoods from the negative consequences of commercial developments.

3.1 Recent Examples of CDA

A good way to get a sense of discourse analysis is to familiarise yourself with examples in the fields of urban studies, housing and planning.

There are useful chapters in a book edited by Lee and Poynton [16] but for a detailed single-case study, see Marston [17].

Analysing housing policy

Marston's [17] book titled *'Social Policy and Discourse Analysis'* examines broad changes within Australian public housing provision. The particular focus of his study is the Queensland government's state housing authority and the way that market-orientated ideologies informed policies were developed and implemented. As Marston explains, the influence of these market ideologies was significant and shaped the way public housing tenants were viewed by the housing authority as 'customers'. There were also changes in a way that services were delivered to reduce costs and encourage tenants to become self-reliant. Marston's study provides a comprehensive and practical resource for readers interested in understanding both the strengths and limitations of CDA. He makes the important claim that CDA on its own is rarely sufficient to capture the complexity of politics and so it often best deployed in conjunction with other data collection approaches. Readers interested in participatory planning processes and the insights that can be achieved through CDA should take a look at MacCullum's book *Discourse Dynamics in Participatory Planning* [18].

Most CDA publications usually appear in academic journals.

Making sense of contemporary planning reform

An excellent analysis is provided by Gurran and Ruming's titled 'Less planning, more development? Housing and urban reform discourses in Australia' [19]. They adopt a Foucauldian approach to explore how specific narratives about problems in the NSW planning system can be viewed as a 'regime of truth' that influenced government legislation for a new planning system. The research draws on written documents and submissions within a government consultation period on the proposed reforms. The objects of their inquiry are the written submissions made in response to the NSW Green and White Papers that were presented to the NSW parliament in 2013 and the draft Planning Bill submissions provided by business, local government, peak bodies and community groups. Amongst their key arguments is that it ideology and economic interests are not the only drivers of planning reforms. Other influences can be sourced to particular features of the NSW parliament when the government did not have a majority. This meant that the legislation was stalled and community opposition was able to influence decision-making. They also noted how development interests used the submission process to argue that the current planning system was too favourable to community groups opposed to development and that economic growth was at risk unless reform was implemented.

There are other studies that are useful for the insights on Australian urban and planning issues. McGrath [20] undertook a critical discourse analysis to examine 31 local government sport and recreation plans to assist people with disabilities. He [20] shows that the community provision offered equates disability with immobility and so overlooks other forms of disabilities during consultation processes. Darcy [21] deploys a CDA to examine mixed-income housing developments in Australia to show how policy discourse operates as a rationale for this form of redevelopment. In his analysis, he refers to the frequency of particular research categories (such as 'estate' and 'mixed income') to occlude the pernicious outcomes of neoliberal policy interventions. MacCallum and Hopkins [22] examine the changing discourses of city planning in Perth between the years 1955 and 2010. They offer an excellent example of how to chart the way that ideologies shift over a sustained period.

Williamson and Ruming [23] in their article 'Urban consolidation process and discourses in Sydney, unpacking social media use in a community group's media campaign' consider the different tactics that community groups deploy to stymie urban consolidation in Sydney. They explore how terminologies are used and the rhetorical strategies adopted to influence opinion on proposed development in the beachside neighbourhood of Bronte, Sydney. Amongst the data they selected for analysis were: planning proposals; reports by local government; NSW government media releases; newspaper articles and social media data from *Twitter*. Their article provides an example of how statistical software can be used to generate empirical data. In their analysis, they showed how stakeholders sought to gain influence by appropriating each other sources and how the large volume of *Twitter* correspondence led to strain on the planning agencies as they felt they were expected to respond.

Sengul [24] deploys a CDA to chart the influence of right-wing Australian politician Pauline Hanson pronouncements on race. Some of the questions that Sengul asks include: How are social groups referred to? What characteristics and qualities are assigned to certain individuals and groups? What arguments are deployed to reinforce these assignations? Sengul distinguishes between the production and consumption of text. What she terms 'perspectivisation' is an important component of her analysis. She provides insights into the ways that texts are crafted to influence public opinion.

Rayner et al. [25] undertook a textual analysis of 449 articles that appeared in Brisbane newspapers between the years 2007 and 2014. Their research identified some of the most common terms that regularly appear in articles and headlines. Amongst their claims are that social representations that are feature of the mass media provide insights on 'common sense' understandings about what urban consolidation entails. They claim, on the basis of their analysis, that there are competing understandings of urban consolidation that operate across different registers that include housing, sustainable population growth, finance and the politics and planning nexus. Popular conceptions, they note, are often negative and reinforce anxieties about new development.

Our last two examples are articles from the UK. A recent study is provided by Joss et al. [26], they deploy a discourse analysis for evaluating smart cities initiatives using *AntConc* and *Nvivo* softwares to track the occurrences of the term 'citizenship' as a first step for considering whether smart city initiatives in the UK are following instructions to become more civic focussed. They look at the discursive frames that have shaped policy response and the effectiveness of the British standard institution. Munroe [27] considers how house price inflation is reported in the UK. Munroe draws upon the work of Fairclough and Wodak [28] and pays attention to the production of texts the 'silences' within the text. As she writes, 'an important insight of CDA is that in using language the writer implicitly is deciding what *not* to say: a significant part of conducting CDA is to interrogate those meaningful silences and absences. These 'choices' in language will often be unconscious or unthinking, as authors may simply draw on conventional or dominant representations, without any active or deliberate intention to espouse a particular ideological position or attempt to persuade or produce propaganda' (Munroe [27], p. 1090). She also examines the metaphors that are used to depict housing prices by using quantitative software tools such as *Lexus-Nexis*. Amongst Munroe's findings is that the housing market is portrayed in a way that is similar to a person that goes through periods of good health and illness (Munroe [27], p. 1099).

Most CDA research is nation-specific but there are studies that are transnational. Our final example is McArthur and Robin [29]. They examine the term 'liveability' discourse in recent strategic city planning across the globe. They show how 'liveability' in a neoliberal policy setting is operationalised in ways that undermine urban livelihoods rather than improve them. They explain that there are considerable silences in the way it is evoked and they suggest that discourse analysis can not only foreground the problematic features of government discourse but also provide insights as to the strategies that are required to counteract these homogametic constructions. As they argue, a recognition of liveability discursive power can help us put in place a counter-narrative that better connects the inequalities and contested aspects of everyday urban life. They see discourses around urban liveability as a linguistic apparatus that frames policy interventions. In their discussion of the Economist Intelligence Unit's Global Liveability Ranking and Global Liveable Cities Index, they show that these rankings use quantification and metrics to foreground the preferences of mainly privileged urban dwellers. In effect, obscuring tensions and conflicts taking place within these cities.

3.2   Recent Examples of Foucauldian Discourse Analysis

The books by Bacchi [30] and Flanagan [31] provide excellent exemplars of Australian Foucauldian scholarship.

The representation of social problems

Bacchi's book [30] is a broad investigation of social policy. She argues that most investigations are too narrowly construed and fixated on resolving problems. Rather than just consider policy prescriptions and their efficacy, there is a need to extend the analysis to account for the ways that policy problems are framed. Bacchi, like Foucault, argues that policy problems are not out there waiting to be resolved but are constructed as issues that need to be resolved. How problems become manifest is to a considerable extent reliant on people and organisations to identify an issue as a problem. Since there are no immutable underlying problems, the way that issues are represented and discussed shape the type of interventions that follow. She provides five analytical questions for examining social problems. What is the problem represented to be in specific policy?; what assumptions underlie this representation? How did this problem representation become manifest? What is missing or left out when problems are represented? How are 'problems' articulated in policy settings?

Delving into the archives

Flanagan's book [31] offers the reader a rich historical account of post-2nd World War public housing policy in Tasmania. Her Foucauldian analysis consisted of a careful reading of archival records to explore how the representation of public housing changed from being a solution to housing challenges to one that was seen as a cause of poverty and a policy failure. By charting the rise and fall of public housing, Flanagan offers wider insights about the role of the state, and the impact of neoliberal ideology on government itself. She examines the shifting subjectivities and policy objects, for example, broadacre estates, rent and tenants. Her analysis shows how certain shifts in discourse can be traced to the trajectories that shape economic and social life. In taking this approach, Flanagan did not research language use but rather examined the objects of discourse within texts. In other words, she asked what is being talked about?

However, Australian Foucauldian inspired analyses are usually in academic journal articles and report smaller case studies; for example, Kerkin [32] attends to the urban planning issues that arise from sex work in St Kilda, Victoria. She analyses the range of the common representations of sex workers noting in particular the stigma and exclusionary discourses and how these inform decision planning protocols. Barnes et al. [33] explore the ways that 'the creative city' has become a key term in Australian planning discourses. They look at the way that power relations are evident in discourses on what an urban village entails using the example of Wentworth Street, in Port Kembla, Wollongong, New South Wales. Whilst Bonham and Cox [34] draw on Foucault's writings to explore the way that urban cyclists are discussed in health sciences. He argues that those advocating for a segregation between cyclists and other road users inadvertently serves to maintain existing travel norms. Bugg and Gurran [35] examine the planning processes when Islamic school developments are proposed in Sydney. They show how long stranding discourses are often used to undermine arguments for developments. As they explain, discourse analysis can reveal the hidden value systems that inform planning decisions. As is evident from the above examples, a common aim of researchers who use discourse analysis is to expose injustice and inequalities. One explicit study is Porter and Barry [36], they examine the 'contact zones' between Indigenous peoples and urban planning in both Victoria Australia and British Columbia Canada. Porter and Barry [36] show the ways that long-standing discourses maintain territorial, political and administrative structures. Cumulatively, the impact of these discourses undermines the efforts of activists to extend the recognition of indigenous rights.

We can note that discourse analysis is often used in the field of housing studies. Ruming's [37] study examined the Australian Commonwealth Government's 2010 social housing initiative that formed part of the economic stimulus measures in the aftermath of the global financial crisis. His work provides insights on how interviews can be coded and then analysed to establish emerging themes. Amongst Ruming's findings is that homeownership 'solutions' are valorised in Australia making it difficult for welfare agencies to put forward non-commodified housing solutions such as public housing. Another housing specific study is Penfold et al. [38] who deploy a Foucauldian approach in their examination of Australian media representations of the 'tiny house' and their impact on understandings of homeowners. They show

how the tiny house both challenges but also normalises long-standing discourses in respect of nature and consumption.

4   How to Undertake a Discourse Analysis

4.1   Example 1—Covid-19 in the Media

In this section, a newspaper editorial on Covid-19 is provided as an example to show how a discourse analysis might proceed. The methods we have outlined below can be extended to any relevant text as well as visual images and film (see Rose [14] for an excellent overview of these methods).

Context

The Covid-19 outbreak continues to have profound impact on nation states across the globe and the economic, political and social impacts are certain to endure for many years. At the time of writing this chapter, the Australian government had formed a national cabinet comprising of the Prime Minister, Treasurer and Premiers of all eight state and territories to coordinate a governmental response. There has been numerous articles and media commentary claiming either that Covid-19 will necessitate a fundamental reshaping of conventional politics or eventualise as a short interruption of 'business as usual'. For many on the left of politics, there is a hope that Covid-19 will put pressure on policymakers to reduce social and economic inequality. On the right of the political spectrum, there is an anxiety that business may lose influence and government will cave into demands that public provision should be extended. The article below was published as an editorial in a conservative-leaning newspaper, *The Australian* [39] on 4 April 2020, shortly after the Coalition Government announced a package of subsidies costing $130 billion to help businesses and employees manage the impact of the lockdown. The editorial provides an example of one of the ways that powerful vested interest groups seek to influence political debates in order to protect their power and privilege.

Without the benefit of hindsight, it is difficult for researchers to take stock of the current period and so much of the predictions written at the time of Covid-19 pandemic may prove to be wrong. This noted, critical discourse analysis explicitly situates the present moment within an historical setting. In practice, it seeks to show how the arguments (set out in the example below) are linked to longstanding ideological debates circulating in Australia about the role of government and the relationship with its subjects. A starting point to acquire the skills to undertake a critical discourse analysis is to read the editorial example provided below. We suggest you read the article twice over, first as you would read any newspaper editorial and second time around in a more critical way that accords with the approach suggested by Norman Fairclough and other exponents of CDA.

Step 1: Choosing a text(s)

When embarking on a critical discourse analysis the first task is to choose a selection(s) of texts for the analysis itself. The decision must be fully justified so that those who are reading your analysis have a clear understanding of your selection. In the example below, we have selected a newspaper editorial that advances an ideological argument to influence political decision-making during the early stages of what can now be described as an exceptional period in Australian politics.

New normal cannot be this Whitlamesque freak show

Last May, voters rejected the statist, welfare-all-around, social engineering, aspiration-killing, emissions-busting manifesto Bill Shorten was proposing. So here we are, fighting the coronavirus by putting a $2 trillion economy into suspended animation. Australia is being transformed before our eyes. Measures unthinkable weeks ago are flashing past, day by day, as Scott Morrison is doing 'whatever it takes' to 'save lives and livelihoods'. This week's $130bn wage-subsidy plan certainly had a wow factor. The $1500 JobSeeker payment goes to employers who keep employees on their books; it is expected to last for 6 months, preserving six million jobs. The Prime Minister has essentially nationalised half of the country's private payroll. Socialist-lite Mr Shorten, eat your heart out.

That third survival package follows what now look like fiscal baby steps of $17.6bn and $66.1bn. That is a $214bn bill in 18 days, without counting $2.4bn on health spending for pop- up Covid-19 testing clinics and hospital stays; $1.1bn for mental health, telehealth services and domestic violence support; and $3bn for free childcare for almost one million families. In a normal year, Josh Frydenberg spends just under $500bn. The nominally conservative, small government, low-tax side of politics is building a welfare state on a vast scale, benevolently taking over sectors such as childcare and private hospitals. Not surprisingly, rent-seekers, grifters and failing enterprises are popping out of every crevice, let alone the chasm that has swallowed tourism, hospitality and education. Virgin Australia, battered not just by travel bans, wants a $1.4bn bailout. The Morrison government told the air-line to milk its shareholders for an emergency capital injection.

Facing a health and economic crisis without precedent, we are creating a burden of deficits and ballooning debt. Sure, money is cheap, especially for government borrowers. But there is an opportunity cost for the $500bn bill Canberra and the states are likely to carry when Covid-19 has been tamed. The burden of repaying debt will fall on young workers and those still learning. Mr Morrison has said 2020 will be for most Australians the worst year of their lives. He insists these are temporary, targeted and proportionate measures. We take him at his word when he claims the spending tap will be turned off pronto, calling it a 'snap-back' to pre-existing welfare arrangements. But the Prime Minister has not yet sketched out benchmarks, such as the number of infections, which will signal its time to take a frail economy out of hibernation.

One of the follies coming out of this crisis is the idea governments have all the answers. We acknowledge the state has a central role in setting the rules of the game when markets fail, delivering basic services and providing a decent safety net for folks thrust into hardship. Government can do a lot, but it cannot do it all. Society functions best when government steps aside, allowing free enterprise and individuals to flourish. A nation where everyone gets a cheque from Canberra or where our elected representatives and officials control the commanding heights of capitalism is on the road to penury. We freed ourselves from this moribund state paternalism when Bob Hawke, Paul Keating, John Howard and Peter Costello opened the nation to competition and harnessed an aspirational ethos: to start businesses, build wealth and be self-reliant.

There will be some who see this momentary neo-Whitlamist fiscal freak show as the dawn of a new politics, where big government rules. It is true the co-operation and co-ordination of national cabinet are superior on urgent matters compared with the clunky Council of Australian Governments format. Mr Morrison is fond of saying there are no red and blue teams in this new grouping of leaders. Perhaps one benefit from this crisis will be a renewed focus on what is important. Voter trust in our democracy has plummeted during a dysfunctional decade of hyper-partisanship, internal party disorder, obsession with niche rights and legislative gridlock. Covid-19 and its aftermath of high debt, social anxiety, and economic fragility will demand the political class turn its attention to basic problem-solving and long-term preparedness—on national security, vital infrastructure, manufacturing capability, sustainability and prosperity.

Centre-right parties, as they seek to redefine their purpose, must be careful not to embrace the false, if voguish, saviour complex. Australia has to pay its way and earn its living in the world. This is not preordained. Mr Morrison must remember why voters rejected Shortenomics: there is no fiscal magic pudding and certainly no shortcut Pavlova Magic for puffing up wages by decree. Higher living standards depend on a dynamic, high-performance economy. In time, Mr Morrison's exit strategy must be based on revivifying the private economy, not on half-baked progressive new deals. It means taking the shackles off all our companies—reducing taxes, red tape and workplace restrictions—and encouraging them to invest, be innovative and employ and train people. Covid-19 has turned life upside down. The fiscal response is not a permanent condition. The state may be your mate now. When this crisis passes, let us get back on our own feet.

Step 2: Political context

The second task is to provide the reader with a clear summary of the political context in which the editorial was produced. The Australian newspaper is part of Rupert

Murdoch's News International media conglomerate and has for many years adopted a partisan style of journalism that is generally supportive of Conservative led governments. This editorial is typical of the Australian in its demand that governments should prioritise business interests when pursuing economic and fiscal policy and remain vigilant about increasing expenditure on collective welfare provision. The editorial is seeking to alarm its readers about some of the attendant risks associated with government interventions in the midst of the Covid-19 outbreak.

Step 3: Analysis of text

The largest part of any critical discourse analysis is the scrutiny of the text(s) you have chosen. Usually, this requires you to select passages or sentences to demonstrate the way the authors who have produced the text are seeking to engender a response from those who read it. In the example above, there are sentences that seek to raise alarm about the direction of government policymaking in Australia. Consider the opening sentences in the editorial, where it is stated that 'voters rejected the statist, welfare-all around, social-engineering, aspiration-killing, emissions-busting manifesto Bill Shorten was proposing'. Here is an example of what is known as framing, by which the writer is seeking to establish a version of reality for the reader. Here there is no attempt to be objective about the opposition party 2019 electoral campaign, which saw the coalition government return to power with a majority of 2 MPs.

Following on from these passages, the editorial asserts that government expenditure is on such a large scale that 'Australia is being transformed before our eyes' and 'the Prime Minister has essentially nationalised half of the country's private payroll. Socialist lite Mr Shorten, eat your heart out'. We can see the attempt is being made to establish an equivalence between the PM's action and the policies ascribed to the former Labor leader Bill Shorten. The style of writing is what CDA analysts refer as genre and in this example, it is polemical with no attempt to appear balanced when claims are being asserted. Editorials and opinion pieces are often more polemical than academic and newspaper reporting, which often try to convey a semblance of objectivity.

The second paragraph in this editorial uses figurative language to embellish the claims being asserted. Consider for example the terms 'fiscal baby steps' 'failing enterprises popping out of every crevice' and 'the crevice let alone chasm that has swallowed tourism, hospitality and education'. This form of language is intended to reinforce the message that readers should be greatly concerned by the new direction in government decision-making.

The fourth paragraph provides a rationale for the arguments developed earlier. Here readers are invited to contrast a society where 'government steps aside, allowing free enterprise and individuals to flourish' with 'a nation where everyone gets a cheque from Canberra or where our elected representatives and officials control the commanding heights of capitalism. In order to assert this binary, the editorial cites the Labor leaders Bob Hawke and Paul Keating along with Liberals such as John Howard and Peter Costello, all of who opened the nation to competition and harnessed an aspirational ethos'. Evoking the names of former politicians is what is commonly referred in CDA as intertextuality i.e. making linkages to trusted sources as a basis to reinforce arguments or boost the authority of the author.

Another important term in CDA is cohesion, this is where authors arrange their sentences in ways intended to evoke a specific response. In the final paragraph, the authors seek to present a choice between those who see COVID 19 as an opportunity to extend a 'fiscal freak show' and initiate the 'dawn of a new politics, where big government rules' and a set of policies that revive 'the private sector companies—reducing taxes, red tapes and workplace restrictions'. Here the order of discourse is arranged to advance an explicit neoliberal ideological position through a denouncement of any possible alternative course of action.

Step 4: Assessment

The final task is to provide an overview assessment of the text, what is referred to by Fairclough as Discursive practice. We can see from the editorial that the author has sought to present an analysis as a self-evident truth What is known as *'truth modality claims'*. The reader is encouraged to think of Australia in stark terms; on the one hand long trusted and sensible policies that protect wealth and encourage self-reliance and; on the other hand autocratic government that is ''statist', welfare-all around- and aspiration killing'. We can see that the *framing* of politics as an ideological battleground is a feature of the contemporary culture, and while there has always been a contest of ideas, it is evident that at this point in time, political divisions have become more polarised and oversimplified by protagonists.

A Foucauldian analysis does not follow the sequential tasks outlined in the above example although it would identify similar issues and connections. The analysis would instead trace the *genealogy* of ideas that are presented in the editorial, discussing, for example, the longstanding opposition to collective provision and the influence of commercial agencies on government policymaking. Another important component of Foucauldian analysis is to draw attention to the silences in the text and ask what is being left out in the text? It is apparent that there is only scant acknowledgement of the important role that government has played during periods of economic crisis and much of the article seeks to disparage nearly all forms of government welfare expenditure. A Foucauldian analysis would identify the self-evident 'truths' that are assumed in this article, noting the portrayal of private sector businesses as productive and the contrast drawn with collective welfare provision, and it would most likely note that valorisation of private enterprise has a long lineage that that can be traced to discourses that first surfaced in the eighteenth century. The Foucauldian approach would also elaborate on the context in which the text itself is situated, perhaps seeing, the global Covid-19 pandemic as a moment of crisis in which discursive shifts may or may not emerge.

4.2 Example 2—COVID-19 Opinion Piece

The transcript below is a media opinion piece [40] authored by the Executive Director of the Tasmanian Property Council during the early days of the lockdown. We have selected it because it provides an example of how business lobbyists seek to influence public opinion. The timing of the article is highly relevant not only because of the impact of Covid-19 on Tasmanian businesses but also because the government was in the process of legislating on planning reforms that would reduce the scope for communities to influence development projects judged by the government to be of State significance.

Talking Point: Get ready to ramp up development

THE COVID-19 threat will be one of the defining moments of a generation, and we need to all pull together in such challenging times. Australia has faced significant economic challenges before and it has always been the strength and resilience of our people that has helped us through. We will get through this and we will rebuild, but it is going to take time, effort, and leadership.

The Property Council of Australia stands prepared and willing to work with local, state and federal governments to ensure that our industry plays its part in supporting jobs, construction and investment throughout this period and into the recovery which will follow.

Commercial landlords should want their tenants to survive and to be in place for the recovery, and while the economic crisis is only in its early days in Australia, building owners are already working with their tenants to support their businesses on a case-by-case basis with short and longer-term solutions.

This may include deferral or abatement of rent, reductions in outgoings, changes in rental frequency and payment plans, and extended grace periods regarding defaults.

The financial support that commercial building owners are likely to provide over the coming months will be very substantial, amounting to many millions of dollars across the economy.

We welcome the Prime Minister's announcement that state and territory governments will work on model laws for commercial and residential tenancies experiencing hardship, and we stand ready to work cooperatively on their implementation. We also welcome clarity about land tax in the big picture but have little to no idea what local government are going to contribute when it comes to rates relief.

Local governments need to consider immediate rates relief for all businesses, including a 100% 12-month rates hiatus for those facing significant hardship, and maintenance of rates at current Average Assessed Value (AAV), not adjusted to increased AAVs that are out of kilter with current property values on the back of the crisis.

The City of Launceston's rates of remission for 6 months shows excellent leadership. All municipalities should follow suit without delay.

Businesses cannot afford to wait for local government, they need decisions and clarity now and it is time for the sector to show leadership and do what the upper tiers of government have done and provide support to ensure that the economy can recover after this crisis recedes.

State and local government must also continue legislative reforms aimed at reducing regulatory handbrakes. Significant progress towards the implementation of a streamlined state-wide Tasmanian Planning Scheme and statutory decision-making frameworks must remain a priority, with the importance of simplified processes key to supporting recovery following this challenging period.

This is the perfect time to reduce red-tape, which continues to slow down the building and construction industry's goal of reaching maximum capacity.

Further, the State Government granted itself a range of sweeping new powers and we need to ensure there is proper scrutiny of the decisions it takes in coming months.

Despite the best of intentions, no government gets it right all the time and that is why Parliament's checks and balances are vital. There is also an opportunity to bring forward infrastructure projects that benefit the state in employment, project readiness and economic recovery.

TasWater, TasNetworks and other government businesses must be urged to begin capital improvement projects straight away with the obvious benefits then available as soon as they are completed and into the future as recovery gathers pace. The Property Council of Australia looks forward to welcoming additional stimulus packages that support our members as they continue to sustain small businesses across Australia. Keeping the economy functioning and implementing measures that preserve jobs must remain key priorities.

COVID-19 has created a level of anxiety in the community that has and continues to undermine confidence. We understand that the pandemic presents a troubling and uncertain future, but our industry remains confident that the resilience and maturity we have developed will assist us through this difficult time.

Brave, swift and decisive steps must be taken. The Property Council of Australia is supportive of clear and concise messaging that maintains a level of confidence across the nation. Further, we reiterate the importance of bringing forward and continuing large-scale infrastructure projects, delivering tax concessions, and following through on legislative reform.

Brian Wightman is the Tasmanian executive director, Property Council of Australia.

Source Mercury Newspaper Tasmania April 7 2020.

A Foucauldian analysis would begin by setting out the specific political context that accompanied the outbreak of COVID 19 in Australia and how an assessment of the economic impact prompted the chief executive of the Tasmanian Property Council to write an article in the Mercury newspaper.

It is evident that a pro-business ideological position is used by the author to make his case for planning reform and support for businesses affected by the Covid-19 lockdown. A Foucauldian analysis would also note the silences in the text, that is, what is not being stated. So, for example, it is worth noting that the author calls for rent abatement for commercial businesses but not tenants in the private rental market. The article provides a clear example of the discursive strategy used in newspaper opinion articles to persuade the reader that the arguments are coherent and justified. Overall, the article sheds light on the ways that lobbyists try to influence public opinion and how the production of texts can reinforce long-standing ideological assumptions about the role performed by businesses and the need to prioritise their needs over and above other groups in society.

5 Challenges and Strategies

In this penultimate section of the chapter, we draw your attention to the main concerns that have been raised by critics of discourse analysis. Some critics have argued that CDA is too narrowly construed and the parameters established by Fairclough limit its potential. This challenge has been addressed by Fairclough et al. [41] in their suggestion that CDA is a mode of analysis and not a template. As they explain, critical discourse analysis is 'a problem-oriented interdisciplinary research movement, subsuming a variety of approaches, each with different theoretical models, research methods and agenda' (Fairclough et al. [41], p. 357). Their point is that researchers have to adapt their methods to the questions they want answering.

CDA and Foucauldian discourse analysis has also been criticised by some researchers for being unrepresentative when texts are self-selected. Critics have in their sights, scholars that focus on one or two documents as a basis to make broad claims. This criticism can be rebutted in two ways. First, if scholars justify their selection and make explicit how their interpretations accord or diverge with similar analysis conducted through other methods and second, when extending arguments provide qualifications and caveats to make them more nuanced. An issue for all researchers who rely on computer-generated software is the ease by which it can capture huge amounts of data with the touch of a keyboard. When using computer search engines, it is essential to make very careful choices as to what data is relevant and discard what is not. When these choices are not made, the data tend to drive the analysis and the findings are not easily discernible.

6 Conclusions

This chapter has introduced discourse analysis as a method for understanding contemporary urban policy and planning practices. It should be clear, from the worked examples, that the aim of the approach is to understand the ways that power is exercised and that discourse is one of the vehicles by which power is enacted and maintained.

The chapter also discussed recent examples of mainly Australian scholarship that has deployed either critical discourse analysis (CDA) developed by the British scholar Norman Fairclough and approaches that draw upon the writings of Michel Foucault. It is the CDA variant that has been most widely used but those interested in a more historical approach to texts have tended to make use of Foucauldian approaches.

The worked examples establish the steps that are usually taken by researchers who deploy discourse analysis as well as the type of findings that can be advanced. Both examples show how language is carefully used to advance ideological claims as to what governments can and should do during a crisis such as Covid-19. The chapter offers some ways that researchers can rebut some of the criticisms that have been made by those who are sceptical of the approach. Those who have used discourse analysis effectively usually see it as a method that is used in conjunction with other methods such as interviews and focus groups. The chapter has also referenced a large number of articles that encompass contemporary practices as well those that offer insights on methods and analysis.

Key points

- Discourse analysis is a method for understanding contemporary urban policy and planning practices. The aim of the approach is to understand the ways that power is exercised and that discourse is one of the vehicles by which power is enacted and maintained.
- Most of the recent examples of mainly Australian scholarship have deployed either critical discourse analysis (CDA) developed by the British scholar Norman Fairclough and approaches that draw upon the writings of Michel Foucault.
- It is the CDA variant that has been most widely used but those interested in a more historical approach to texts have tended to make use of Foucauldian approaches.
- There are specific steps that are usually taken by researchers who deploy discourse analysis . The aim of the approach is to show how language is carefully used to advance ideological claims.
- Researchers can rebut some of the criticisms that have been made by those who are sceptical of the approach. Those who have used discourse analysis effectively usually see it as a method that is undertaken in conjunction with other methods such as interviews and focus groups.
- There are a large number of articles that encompass contemporary practices as well those that offer insights on methods and analysis.

Further information

For those wanting to get further information about discourse analysis, there are a number of useful guides. Amongst the most useful introductory texts on critical discourse analysis are:

- Norman Fairclough, Phil Graham, Jay Lemke & Ruth Wodak, (2004) 'Introduction', *Critical Discourse Studies*, 1 (1) 1–7
- TeunVan Dijk, (2015) 'Critical discourse analysis' in D. Tan- nen, H. Hamilton, & D. Schiffrin (Eds.), *The Handbook of Discourse Analysis*, Chichester: Wiley Blackwell, 352–371.

and for a Foucauldian inspired discourse analysis, readers should take a look at:

- Niels Andersen, (2003) *Discursive Analytical Strategies*, Bristol: Policy Press.

Acknowledgements Some of the ideas set out above draw from an earlier chapter titled 'Discourse analysis' that was published in M. Walter (ed) (2019) *Social Research Methods: An Australian Perspective*, Melbourne: Oxford University Press (Revised—4th edition) pp. 316–340.

References

1. Fairclough N (1992) Discourse and social change. Polity Press, Cambridge
2. Fairclough N (1995) Critical discourse analysis. Longman, London
3. Fairclough N (2003) Analysing discourse: textual analysis for social research. Routledge, London
4. Fairclough N (2013) Rethinking critical discourse analysis: the critical study of language. Routledge, London
5. Wodak R, Meyer M (2009) Methods of critical discourse analysis. Sage, London
6. Van Dijk T (2015) Critical discourse analysis. In: Tannen D, Hamilton H, Schiffrin D (eds) The handbook of discourse analysis. Wiley Blackwell, Chichester, pp 352–371
7. Fairclough N, Graham P, Lemke J, Wodak R (2004) Introduction. Critical Discourse Stud 1(1):1–7
8. Foucault M (1971) Madness and civilisation. Routledge, London
9. Foucault M (1974) The order of things: an archaeology of the human sciences. London: Tavistock. 10
10. Foucault M (1977) Discipline and punish. Penguin, Harmondsworth
11. Foucault M (1980) Power/knowledge: selected interviews and other writings 1972–1977. Harvester, Brighton
12. Dean M (1999) Governmentality, power and rule in modern society. Sage, London. Fairclough N (1989) Language and power. Longman, London
13. O'Farrell C (2005) Michel Foucault. Sage, London
14. Rose G (2016) Visual methodologies. Sage, London
15. Andersen N (2003) Discursive analytical strategies. Policy Press, Bristol
16. Lee A, Poynton C (eds) (2000) Culture and text: discourse and methodology in social research and cultural studies. Allen and Unwin, Sydney

17. Marston G (2002) Critical discourse analysis and policy orientated research. Housing Theor Soc 19(2):82–91
18. MacCallum D (2009) Discourse dynamics in participatory planning. Ashgate, Aldershot
19. Gurran N, Ruming K (2016) Less planning, more development? Housing and urban re-form discourses in Australia. J Econ Policy Reform 19(3):262–280
20. McGrath R (2009) A discourse analysis of Australian local government recreation and sport plans provision for people with disabilities. Public Manag Rev 11(4):477–497
21. Darcy M (2010) De-concentration of disadvantage and mixed income housing: a critical discourse approach. Housing Theor Soc 27(1):1–22
22. MacCallum D, Hopkins D (2011) The changing discourses of city plans: rationalities of planning in Perth 1955-2010. Plann Theor Practic 12(11):485–510
23. Williamson W, Ruming K (2017) in their article 'urban consolidation process and dis-courses in Sydney, unpacking social media use in a community group's media campaign. Plann Theor Pract 18(3):428–445
24. Sengul K (2019) Critical discourse analysis in political communication research: a case study of right-wing populist discourse in Australia. Commun Res Pract 5(19):386–392
25. Raynor K, Matthews T, Mayere S (2017) Shaping urban consolidation debates: social representations in Brisbane newspaper media. Urban studies 54(6):1519–1536. Schon D (1983) The reflective practitioner. Ashgate, Aldershot
26. Joss S, Cook M, Dayot Y (2017) Smart cities: towards a new citizenship regime? A discourse analysis of the British smart city standard. J Urban Technol 24(4):29–49
27. Munroe M (2018) House price inflation in the news: a critical discourse analysis of newspaper coverage in the UK. Housing Stud 33(7):1085–1105
28. Fairclough N, Wodak R (1997) Critical discourse analysis. In: Van Dijk T (ed) Discourse as social interaction. Discourse studies: a multidisciplinary introduction, (2nd ed). Sage, London, pp 258–283
29. McArthur J, Robin E (2019) Victims of their own (definition of) success: Urban discourse and expert knowledge production in the Liveable City. Urban Stud 56(9):1711–1728
30. Bacchi C (2009) Analysing policy: what's the problem represented to be?. Pearson, Frenchs Forrest NSW
31. Flanagan K (2020) Housing, neoliberalism and the archive: reinterpreting the rise and fall of public housing. Routledge, New York
32. Kerkin K (2004) Discourse, representation and urban planning: how a critical approach to discourse helps reveal the spatial re-ordering of street sex work. Aust Geogr 35(2):185–192
33. Barnes K, Waitt G, Gill N, Gibson C (2006) Community and nostalgia in urban revitalisation: a critique of urban village and creative class strategies as remedies for social 'problems'. Aust Geogr 37(3):335–354
34. Bonham J, Cox P (2010) The disruptive traveller?: A Foucauldian analysis of cycleways. Road Transp Res J Aust N Z Res Pract 19(2):42
35. Bugg L, Gurran N (2011) Urban planning process and discourses in the refusal of Islamic Schools in Sydney, Australia. Aust Plann 48(4):281–291
36. Porter L, Barry J (2015) Bounded recognition: urban planning and the textual mediation of Indigenous rights in Canada and Australia. Crit Policy Stud 9(1):22–40
37. Ruming K (2015) 'Everyday discourses of support and resistance: the case of the Australian social housing initiative. Housing Theor Soc 32(4):450–471
38. Penfold H, Waitt G, McGuirk P (2018) Portrayals of the tiny house in electronic media: challenging or reproducing the Australian dream home. Aust Plann 55(3–4):164–173
39. The Australian Newspaper (2020) 'New normal cannot be this Whitlamesque freak show' Editorial 4th April. https://www.theaustralian.com.au/commentary/editorials/new. Accessed 6th Apr 2020
40. Whiteman B (2020) 'Get ready to ramp up development' Mercury Newspaper 7th April. https://www.themercury.com.au/news/opinion/talking-point-get-ready-to-ramp-up-dev elopment/news-story/ca590f4b090865bb4484ee2d2974bb53. Accessed 9th Apr 2020

41. Fairclough N, Mulderrig J, Wodak R (2011) In: Van Dijk T (ed) 'Critical discourse analysis' in 'discourse studies: a multidisciplinary introduction. Sage, London, pp 357–78

Evaluation Research

Paul Burton

Abstract Evaluation research is an important part of the broader field of urban analysis. It draws on many of the same techniques and approaches used by other types of research about urban areas and processes, including social, economic and environmental science methods, but uses them to make judgements or evaluations. These evaluative judgements are not based on the preferences or priorities of the evaluators (although they invariably have them), but on criteria established by others. These criteria might be the aims and objectives of an urban policy intervention or the assumptions underpinning a land-use planning scheme. This chapter traces the development of different conceptions and approaches to evaluation, from experimental and constructivist through to realist approaches. It looks also at the particular challenges faced by evaluation researchers looking to judge the success of complex, area-based initiatives which are still a mainstay of much contemporary urban policy. The chapter looks also at the politics and ethics of evaluation research, including the challenges of 'speaking truth to power', and recommends the continuous development of the craft skills of evaluation research alongside an ongoing commitment to the fundamental principles of good research design.

1 What is Evaluation Research?

As earlier chapters have discussed, urban analysis covers a multitude of perspectives and practices with a common focus on phenomena and processes that have an urban dimension. Of course, this does not necessarily help us in defining what we mean by 'the urban' with any precision as many processes such as job creation and loss, housing and community development or travel behaviour have fundamental characteristics that may or may not vary according to their urban setting. Likewise, many phenomena that we might wish to study in an urban setting, such as unemployment, social isolation or anti-social behaviour, can be found also in suburban or rural settings. But, for the time being, let us put to one side the challenge of differentiating

P. Burton (✉)
Griffith University, Gold Coast, QLD, Australia
e-mail: p.burton@griffith.edu.au

© Springer Nature Singapore Pte Ltd. 2021
S. Baum (ed.), *Methods in Urban Analysis*, Cities Research Series,
https://doi.org/10.1007/978-981-16-1677-8_10

urban phenomena from those generated by or experienced in other types of area and focus instead on evaluation research as a vital element of urban analysis, for while the phenomena in question might vary, the means of analysing them and evaluating policy responses to them is not significantly different.

However, definitional challenges exist also when we consider the field of evaluation and evaluation research, not least in exploring the specificity of the main techniques of evaluation research *vis a vis* social and economic research more generally. What can be said about any form of evaluation, which inevitably requires research in one form or another, is that it is judgemental. This is not to say that moral judgments are made—although they might be—simply that evaluation involves assessing something against a set of existing criteria. As the renowned evaluation theorist, Michael Quinn Patton [1, p. 1] observesd in a recent critical summary of the field,

> It's all about criteria. Criteria are the basis for evaluative judgment. Determining that something is good or bad, successful or unsuccessful, works or doesn't work, and so forth and so on requires criteria

In other words, while urban analysis is a broad field it includes the similarly broad field of evaluation research, which is an unavoidably judgemental exercise in which degrees of success or failure are judged against criteria. In the urban realm, those criteria could be a plan or strategy, they could be a new planning process, or they could be more tangible outcomes seen on the ground—a new suburb or a park, for example. Evaluation in this sense relates typically, but not exclusively, to interventions that are expected to have certain desirable outcomes, but these interventions can originate from public or statutory bodies such as local or state governments or from private developers be they large corporations or individuals.

Some have argued the case for 'non-judgemental evaluation' [2] but this has always seemed (to me at least) to be a *non sequitur*, a proposition that does not follow logically from its premises. An advocate for non-judgemental evaluation might claim to be providing only a description of the outcomes and impacts from an intervention, while leaving the conclusive judgement to others. But this suggests the initial descriptive work is only part of the evaluation process, an important part to be sure, which is not complete until someone makes a judgement call. It is worth noting that this position is similar to the philosophical distinction made between facts and values and to the practical distinction that underpins the notional relationship between civil servants or state bureaucrats and their political masters. In this case, politically neutral civil servants provide advice to politicians who make decisions based on their clearly articulated political values.

There is, however, a well-established viewpoint within the literature on the theory and philosophy of evaluation [3] that says we should not begin with a set of clearly stated objectives or goals and examine the extent to which they are realised in practice. Rather, we should look at what happens in practice and then try to deduce a set of objectives that would make sense of that practice. This is a logical approach to take, especially if you prefer an inductive rather than deductive approach to scientific exploration. However, it struggles with one of the same practical challenges of inductive methods, namely that it can be extremely difficult (some would say impossible) to clear one's mind of preconceptions before collecting and analysing empirical data and even to know where to focus one's analytical gaze at the start of this process.

In this chapter, I assume, therefore, that the most productive way to plan and carry out a piece of evaluation research is to look at the stated objectives, aims or goals of an intervention and then to think about how you might go about assessing the extent to which they have been realised. In the course of this and perhaps in conclusion, one might argue that the stated objectives were mainly symbolic [4] and there was never any genuine intention of carrying them out or manifestly insufficient resources were devoted to their implementation, in which case a more covert set of objectives might be described. But the essence of evaluation research involves judging something by the standards it sets itself.

To summarise, the broad field of urban analysis includes evaluation research, or research carried out as part of an evaluation. Much research in the urban field involves the evaluation of urban policy measures, such as area-based initiatives which we will look at in more detail later. But it includes also research carried out as part of the development of urban policy and in the implementation of planning policy, so that planners are able to judge whether or not a development proposal meets the requirements and intent of a planning scheme and can be recommended for approval.

This chapter explores some of the key methodological, conceptual and practical challenges of doing evaluation research in urban settings and about urban processes and phenomena. It does not provide detailed instructions for how to go about designing and conducting this type of research as there are many excellent books on these practicalities, listed at the end.

Finally, it is important to acknowledge the significant differences that can exist between urban analysis or research produced by academics for other academics and urban research produced with the primary intention of being useful to policy-makers and practitioners. Of course, academic researchers can choose whether to direct their work mainly at fellow academics or at practitioners and some choose to serve both audiences, and researchers beyond the academy can likewise choose to present their urban analytical work and/or evaluation research, undertaken initially for practitioners and policy-makers, to academic audiences. There is a growing body of work that examines the relationship between research for these different audiences, exploring the motivations of researchers and their ability to present critical conclusions. While some of this new work offers respectful and thoughtful insights on the challenges of moving between these worlds [5–7], there is another body of work that claims to be critical but is characterised by a tendency to political naiveté and academic piety built on unsophisticated conceptions of the relationship between research, policy and politics [8].

2   Methodological Challenges

If much of the substance of urban analysis and evaluation research involves questions of what works, to what extent, and at what cost, then issues of causation are central. Sometimes in evaluation research, we talk of theories of change or program logic when referring to underlying causal mechanisms, but sometimes we find it difficult to identify them at all or we struggle to grasp their complexity and fall back on notions of 'black boxes'. These metaphors allow us to deal with inputs, outputs and outcomes, without understanding the processes by which inputs are turned into outputs that have outcomes. For example, I have a number of black boxes that take inputs of electricity and data stored on a compact disk and turn it into music that I can hear and enjoy. While I know that without those inputs, I would not be able to listen to music from that black box (which might of course be white or silver), I have little or no idea of what goes on inside it: how data are extracted from the disk, turned into music and broadcast to me. I might be able to assess the quality of these inputs (the reliability of the electricity supply, or the significance of scratches on the disk) and know that they are likely to affect the quality of the output and hence my enjoyment of it, but, again, I know little about why a scratch cannot be dealt with by the CD player.

Some evaluation research deals with complexity and a lack of understanding of causal mechanisms by using a similar black box metaphor and focussing instead on more tangible and measurable inputs and outputs. I was involved in various evaluations of UK urban policy initiatives in the 1980s and 1990s, commissioned by the government departments responsible for them, in which the emphasis was almost entirely on the measurement of inputs and outputs, with only occasional reference to outcomes and rarely any consideration of causal mechanisms. As noted in the introduction, many of these took the form of area-based initiatives in which selected areas would receive additional funds and sometimes enhanced powers to alleviate local manifestations of poverty, deindustrialisation, environmental degradation and poor local public services. As part of the evaluation of these programs, we would scrutinise budgets, check that money allocated to local groups to deliver services such as childcare, environmental improvements or job training was properly spent and then monitor certain outputs. These might include the number of new childcare places available, the number of trees planted, or training sessions delivered. Sometimes, our measurements would be rather more sophisticated and we would track whether the childcare places enabled parents to look for work or have the time to attend job training sessions, or whether the saplings planted ever managed to grow into established trees or whether the job training sessions helped people get jobs. Because each of these measures in themselves provided only a very partial picture of the anticipated transformation of an area, a summative assessment was also important but rarely called for or delivered. As we will explore in more detail below, there are considerable methodological problems in establishing robust causal connections between intervention measures and outcomes, but that is no good reason for not attempting to do so. This more rounded evaluation would also scrutinise the assumed causal connections and mechanisms that lead policy-makers to assume (in this case) that a lack of childcare places inhibit the labour force participation of parents of small children, or that planting trees helps restore a sense of pride and wellbeing into residents or that job training enhances employability. Without this focus on causal mechanisms, we are not well placed to claim that we have good theories of causation, such that we can say with some confidence that a combination of policy intervention X, Y and Z produced this particular outcome for this place. As we will consider below, a variety of factors often combine to make this type of summative evaluation very difficult to carry out.

But first, an important digression into the philosophy of knowledge and science and a relatively recent development that has had a profound impact on the design of robust evaluations. While philosophical realism can be traced back to Plato, it was not until the 1960s that works now considered seminal in the field began to appear— Rom Harré's Introduction to the Logic of the Sciences [9] and Roy Bhaskar's A Realist Theory of Science [10] being good examples. Critical realism is in essence the belief that causal mechanisms exist and account for why certain things happen. The 'things that happen' are real and not simply figments of our imagination and can be captured or described empirically. This position developed and extended its reach into the field of evaluation for a number of reasons, not least because of a growing disquiet with on the one hand the excesses of constructivism, especially when combined with impenetrable postmodern jargon, and on the other a rejection of the simplistic and a theoretical assumptions of empiricism.

In 1997, Pawson and Tilley's Realistic Evaluation [11] provided the first comprehensive and compelling summary of how to apply the principles of critical realism in the evaluation of public policy measures. The opening chapter of their book presents (in 28.5 pages) a brief history of evaluation over the preceding 30 years, including what had become known as 'the paradigm wars' between advocates of experimental approaches, pragmatists, constructivists and latterly, realists.

The experimental approach is rooted in a common-sense notion of causation that relies on comparing similar or identical groups (perhaps of people or places), one of which receives an intervention (an expedited planning regime for example) while the other does not. By measuring the volume and throughput of development before and after a given period of intervention (perhaps one or two years in this case), we then assume that any difference is attributable to or caused by the intervention. While this approach can be made to work reasonably well in situations like a laboratory where all the possible variables or characteristics of the two groups can be controlled, in the field or 'real world' this is not only more difficult, it is perhaps impossible. Some have argued that policy reforms or interventions that are selective rather than universal allow for comparisons that are close enough to the assumptions of classical experimental design to be worthwhile [12–14] and more recently a number of governments, including in Australia and the UK have established new policy research teams that use experimental approaches and behavioural economics to provide more robust empirical evidence to policy-makers.

In the face of a growing volume of evaluation research that struggled to find any definitive answers to the question of 'what works?', a new strand of pragmatism emerged among evaluators. This made the assumption that the practice of poli-cymaking and policy implementation is complex and messy rather than rational and coherent and, therefore, evaluation research needed to adopt a more nuanced approach, which recognised that policies and interventions will change over time and that evaluation research will be only one factor in this turbulent environment. It also recognised that the value placed on research and its conclusions is a messy process that was, to some extent, a product of political values and preferences as much as any claim to objective truth. In other words, a piece of evaluation research might be more influential in policy circles if its conclusions aligned with the priori-ties of leading politicians and bureaucrats regardless of its methodological rigour or epistemological purity.

This movement away from absolute notions of truth and objectivity reflected debates in the much wider field of epistemology and the philosophy of science and social science. But more importantly, it reflected a recognition and acknowledgement that evaluation research is unavoidably part of a political process and that the power relationships between those who commission and those who conduct an evaluation are significant. Critical commentators argued that the principle of 'whoever pays the piper, calls the tune' must be recognised as it calls into question some claims to the maintenance of professional standards and objectivity. We return to this below when considering the challenge of 'speaking truth to power'. During the 1980s and 1990s, the constructivist conception of evaluation held sway [15] wherein evaluators nego-tiate or construct what might now be termed an assemblage of perspectives and views held by the full range of stake-holders in any intervention or policy regime. Critics of this position point to the inequalities of power held by this range of stakeholders and its consequences for trying to synthesise and reconcile these views and opinions and even Guba and Lincoln [15, p. 45] recognised some of the conceptual challenges of this approach when they said,

> Evaluation data derived from constructivist inquiry have neither special status nor legitima-tion; they represent simply another construction to be taken into account in the move towards consensus.

A variant of this recognition underpinned the development a more pluralist conception of evaluation popularised by Rossi, Freeman [16]. This set out a more catholic stance in which it was assumed to be possible, indeed necessary, to accept and combine a variety of approaches into a more comprehensive frame. As Pawson and Tilley [11, p. 24] put it,

> One can imagine the attractions of a perspective which combines the rigour of experimen-tation with the practical nous on policy making of the pragmatists, with the empathy for the views of stakeholders of the constructivists.

By avoiding the thorny question of whether fundamentally different approaches to research and indeed conceptions of knowledge can be combined, the pluralist approach did help evaluators by suggesting that in practice, different aspects of an intervention might be best understood through different research approaches, even if coming to a conclusive synthesis remained a practical and conceptual challenge.

Finally, we see the emergence of theory-driven evaluation in the work of Chen and Rossi [17] and Chen [18], which seeks to explain not just whether an intervention appears to work, but why it does so. An important element of this approach, which contrasts with one of the fundamental principles of the classical experimental approach, is the treatment of variables that might explain differences in outcomes. While experimentalists typically look to isolate the one factor or variable that might explain success (or failure), the more modern theory-driven approach recognises that a number of variables might help explain why an intervention works well in one place at a particular moment in time, but not elsewhere or at a different moment.

This then served as a foundation on which Pawson and Tilley built their conception of realistic evaluation. They start with a notion of theory that contains three elements, linked in ways that help us understand the processes of causation. The first, causal mechanisms (M) are plausible accounts of why something might produce an effect because of the connections between them. The second are contextual factors (C) that influence the extent to which these causal mechanisms come into play and the third are the outcomes (O) produced by this combination of causal mechanisms and contextual factors. They summarise this as:

$$M + C = O$$

Consider this example from urban policy: a local authority constructs a small factory and workshop space in the belief that it will stimulate (cause) local economic development through the establishment or growth of small businesses. However, if the rents charged or lease terms are not sufficiently attractive (contextual factors) then the objective or outcome of local economic development will not be realised. From this, we should not necessarily conclude that the theory of building small factory and workshop units to stimulate local economic development is flawed, but rather that to work effectively it requires some other contextual factors (such as suitable rents and lease terms) to be in place. Hence advance factory building might work very well in one place, but not in another. An evaluation based on one case study would struggle to reach this conclusion, while one that involved a lot of similar interventions in different settings (contexts) would enable this more sophisticated conclusion to be reached.

3   Evaluation Research and Urban Analysis

Having explored some of the difficulties encountered in conceptualising evaluations in general, we can now focus on urban policy evaluation, bearing in mind my claim that the distinctiveness and specificity of the 'urban' remain unclear and contested. And we need to be wary of suggestions that evaluation research carried out as part of a broader process of urban analysis draws on a separate set of techniques and approaches. But what is the domain of urban analysis?

Urban policy initiatives have for many years, in Australia and elsewhere, involved place-based interventions. This approach has many features that are attractive to policy-makers: it allows resources to be targeted to particular areas where problems are greatest and especially inter-connected, and it enables experimental approaches to be explored and modified before being applied more widely. But there are problems as well, not least that it requires a high degree of institutional coordination and cooperation in practice as well as in rhetorical commitment and it has to cope with the challenges associated with the notion of the ecological fallacy. This recognises that characteristics that apply at the group, or in this case neighbourhood scale, will not apply to all or even the majority of individuals living in that area. In the case of a place-based urban policy measure that helped every firm in a particular neighbourhood on the grounds that overall firms were struggling, some might not be struggling but receive this assistance, nevertheless. This undermines the claim that area-based initiatives are especially effective at targeting scarce resources where they are most needed.

Another more serious set of challenges to researchers responsible for evaluating this commonplace form of urban policy was described many years ago by Robson and Bradford [19] who identified a number of conceptual problems they called 'the six Cs'.

The counterfactual problem is one the most difficult and asks, 'what would have happened without this policy intervention?'. In laboratory settings, the traditional way of addressing this problem is to use an experimental design in which a sample of people who are otherwise similar are randomly allocated between two groups, one of which receives the intervention (a drug perhaps) while the other receives a placebo or dummy drug. Because the recipients are otherwise similar then any differences in experience or outcome are attributed to the effect of the drug. While this research design is seen by many as the gold standard in rigorously evaluating impact or effect, it has also been subject to sustained and considerable criticism. Although theory and practice are constantly evolving, it is still difficult to select 'policy on' and 'policy off' areas in the neighbourhood or suburb scale that can be managed in ways that achieve the requirements of a classic experiment.

Confounding factors include other measures that might produce similar effects to those under investigation in an urban policy evaluation. For example, in a particular location, there might be a policy measure that provides exemptions from certain business taxes (rates, payroll tax, etc.) for local business in order to support and stimulate their establishment or expansion. During the course of this program, interest rates set at the national level might decrease significantly and affect the operational costs of all businesses, including those in the intervention location. If local businesses are seen to be doing well or better than usual in this period, it can be difficult if not impossible to attribute this to the local tax concession or to the national interest rate change. Unlike in a laboratory setting, it is rarely possible to control any possible confounding variables in order to isolate the effect of the intervention in which we are interested.

Contextual factors describe the peculiarities or specificities of particular settings and are important in helping explain why general descriptions or explanations might vary because of these peculiarities. This distinction has been developed and applied in the most sustained and sophisticated manner by proponents of realistic evaluation such as Pawson and Tilley (1997) and more recently by members of the RAMESES projects (http://www.ramesespro-ject.org/Home_Page.php). As we noted above, the corner-stone of realistic evaluation is the formula: Outcome = Mechanism + Context, which represents both a critique and an extension of the basic question that under-pinned evaluation research undertaken as part of the evidence-based research move-ment, namely 'what works?'. The basic question for realist evaluators became, 'what works in which circumstances and for whom?'. In other words, locally variable circumstances or contexts will affect local outcomes, even if there is a more general underlying causal mechanism at work. Robson and Bradford [19] illustrate this by noting that in selecting local places or neighbourhoods for an urban policy interven-tion, there will be significant differences in local context that must be considered when evaluating the impact or success of that intervention. These differences might include the social composition of the area and how it has changed over time, its history of economic development and the tradition of community development and political engagement.

Contiguity challenges relate to what are also known as 'spillover' and 'shadow' effects. Spillover effects, as their name suggests, occur when the benefits (or indeed the negative effects) of an intervention are not confined to the target area but spill over into surrounding areas. This can present challenges to the evaluation researcher who has to decide how far afield to go in search of measurable impacts, be they positive or negative. Shadow effects operate in the other direction of causality and can be similar to the confounding effects described above, although they typically are more localised. For example, a major employer located just outside a target area might close and have a consequential impact on small firms within the target area that previously supplied goods to the major firm and services to its workers.

These contiguity challenges, along with those presented by variable contextual factors, are associated with all area-based policy initiatives because they rely on drawing boundaries that include and exclude places and people, sometimes on a fairly arbitrary basis.

Combinatorial problems arise when slightly different mixes or combinations of urban policy measures are applied in the areas targeted for intervention. Because these combinations are not typically described or measured with any degree of precision, it is difficult to attribute any variability of overall success to differences in combination or to other factors. For example, while 10 areas might be selected for an urban intervention 'package', each of these 'packages' might involve a different mix of measures, such as rate relief for local firms, employment subsidies for hiring local workers and environmental improvements. One might offer 50% rate relief for twelve months, while another offers 30% relief for 36 months; some might rely on tree planting to improve the local environment while others deploy a rapid response graffiti cleaning service. In this situation, while all of the targeted areas enjoy the similar status of being an Urban Policy Pilot Area (for example), each is implementing this program in slightly different ways and the evaluation researcher is faced with yet another significant problem of attribution.

Changes over time present the final challenge to evaluators. It is commonplace to argue that place-based urban policy interventions designed to solve complex and multi-faceted problems cannot be expected to solve significant problems in the short term, which might reasonably be taken to mean 2–3 years. Many have argued that these interventions must be left in place for decades if they are to have any chance of achieving profound and transformational improvements [20]. If this long-term commitment is achieved, and it must be said that most programs of this type are either abandoned or radically transformed after a few years, then the evaluator has to have a framework that takes account of changes over time. These changes will be complex and difficult to analyse. For example, some measures might be implemented immediately as they are seen as foundational for the success of other measures and then phased out having done their job. For example, environmental cleanups or crime prevention measures might be seen as a necessary precursor to the introduction of more direct economic stimulation measures, but measures that can be wound up or wound down once they have achieved short term impacts. Or, they might be abandoned because they are judged (by interim evaluation) to have not been successful. Again, there is often significant variation in the pattern and timing of these changes between different target areas and for good reason. What works in one area, might not work so well in another. Capturing this variation and building it into a comprehensive analysis of what works is difficult but essential to any rigorous evaluation.

While Robson and Bradford [19] identified these methodological challenges in designing rigorous evaluation frameworks for complex area-based initiatives some years ago, they are still pertinent today, not least because so many urban policy interventions retain their focus on select and closely defined localities. Interestingly, some of the measures brought into counter the spread and impacts of the COVID-19 pandemic have again drawn attention to the logic of spatial targeting and the boundaries used in particular instances. While popular debate rarely uses the term, it is often about the ecological fallacy whereby the assumed characteristics of a given area or suburb are taken (erroneously) to apply to everyone living in that area. We are increasingly aware that spatial targeting should be used with great caution as part of any urban policy intervention and that new datasets containing much more individually tailored information might be more appropriate in focusing policy measures on those who most need them.

4   Evaluation Research as Part of Urban Planning

In addition to the plethora of research carried out as part of urban policy development and evaluation, we should not forget that statutory land use planning as practised around the world, typically involves a significant research and evaluation component, even if it is not always seen in this light. Planners carry out research on urban conditions as part of the planmaking process and incorporate many assumptions, for example about future populations, that are subject to evaluative scrutiny over time. More significantly, development proposals and applications are assessed or evaluated against increasingly complex sets of criteria that are enshrined in local planning schemes and strategic plans. While not often recognised as such, this is probably the most significant form of evaluation research that is undertaken on a day-to-day basis by planners. Development proposals are assessed in terms of their environmental and social impacts and recommended for approval or rejection against the criteria contained in the planning scheme. This typically requires the planner to either conduct the assessment themselves or to commission and use a specialised assessment from another expert: in ecology, hydrology, soil mechanics, acoustic engineering and so on.

There is, however, another way in which planners carry out evaluation research. Planning theory sometimes distinguishes between procedural and substantive theory, in other words between theory about the processes adopted by planners, such as comprehensive or inclusive approaches when going about their business and the substantive issues they focus on, such as community development, creating vibrant public realms, ensuring transport systems work well or maintaining housing standards. In this respect, evaluation research in planning tends to focus on the substantive interventions of the planner and there has been a dearth of research that evaluates some of the key procedural claims and assumptions about planning, such as the benefits of greater public participation [21, 22].

5   The Politics and Ethics of Evaluation Research

Because evaluation research can never be simply an exercise in the application of technical skills, but inevitably requires judgements to be made and values to be applied, it is a process imbued with politics. This is not to say that evaluation research is necessarily a partisan activity, in which 'experts' associated clearly with different political groups or parties compete to have their assessment accepted—although this of course happens. Rather, it suggests that judgements involve choices, and these have consequences that affect different groups or individuals in different ways. Understanding and appreciating that one is working in a political environment should lead the accomplished evaluation researcher to be aware of the choices they make, including those that appear to be simple methodological choices, and their consequences. This is not to say that important principles should be compromised, but that practitioners may well be confronted with challenges to their principles and should be self-conscious in upholding them or otherwise.

In the remainder of this section, we explore some of these challenges through three distinct but related questions:

- Can researchers be objective and neutral in their work?
- How should we learn from mistakes?
- Can and should evaluation researchers speak truth to power?

5.1 Can Researchers be Objective and Neutral in Their Work?

While absolute objectivity and neutrality—or perhaps impartiality is a better term—might be unattainable, evaluation researchers can choose to make a commitment (at least to themselves) to conduct an evaluation that could in principle conclude with a position they do not like or prefer. One could argue that if an evaluator is unwilling to make this principled commitment, they should not accept the commission of that evaluation. Similarly, if evaluation researchers have reservations about the ways in which those commissioning an evaluation might use their work, they should consider carefully whether or not to accept or bid for the commission. These reservations might include having control over the design of an evaluation, having access to all relevant data, being able to present conclusions honestly reached, having some control over how the results are published and presented, and being able to publish results separately—perhaps in academic journals.

As with any commission, the more clarity that is provided at the outset about these 'terms of engagement', the better. But disputes might still arise around one or more of these aspects which will test the integrity and good faith of all parties. Developing a thoughtful and comprehensive contract helps in these processes and can be seen as a sign of evaluation becoming more professional in its approach.

Rossi et al. [23, p. 404] speak of evaluation becoming a more 'professionalized' field, even if it has not become a profession in the widely accepted sense of the word. Part of this process involves being able to identify and adhere to certain standards when conducting an evaluation. While there is no universally accepted set of standards that must be adopted by any practising evaluator or evaluation researcher, common standards typically include:

- A commitment to systematic, rigorous and perhaps evidence-based work;
- A promise of a degree of competence and experience in conducting evaluations that are systematic and rigorous;
- An expectation that evaluators are honest in their work, not falsifying data, wilfully misinterpreting it or hiding data that does not support the conclusions reached;
- A promise to engage with program managers, clients and other stakeholders in a respectful manner;
- A commitment to framing evaluation in relation to some notion of the public interest or common good.

Of course, each of these principles can be and are subject to criticism. For example, there is now an abundant literature that takes issue with the often simplistic assumptions of evidence-based policymaking and practice (eg., Shaxson and Boaz [7] and Burton [24]), especially in its earliest manifestations. It can be difficult to disentangle concerns with the conduct of an evaluation from dislike of its findings and conclusions and respect for program workers might be difficult to maintain if an evaluation uncovers widespread fraud or egregious behaviour. As in some of our earlier discussion, these principles are in many respects not specific to evaluation practitioners, evaluation researchers or urban analysts. They are fairly universal and indeed the United Nations Educational, Scientific and Cultural Organization (UNESCO) has published an updated Recommendation on Science and Scientific Researchers, which sets out a series of similar statements about how research should be undertaken [25]. To drive the implementation of this, a global network—the Responsible Research and Innovation Networked Globally (RRING)—has been formed with the following mission:

> We are a coalition that has activism at its core. We seek to make research and innovation systems everywhere more responsible, inclusive, efficient and responsive as an integral part of society and economy. (https://rring.eu/)

But as many program evaluators discover through their own work, there can be a significant gap between the statement of program goals and objectives and their achievement in practice and it remains to be seen whether or not statements such as these succeed in raising the standards of actually existing evaluations or remain as mere statements of good intent.

One of the biggest and longest-standing challenges to the integrity of evaluation research lies in the commitment to speak truth to power. While Aaron Wildavsky's [26] classic text, Speaking Truth to Power: The art and craft of policy analysis popularised the term at a time when policy studies were emerging as a distinctive discipline, the expression (parrhesia in Greek) first emerged in the writings of Euripides and was an essential component of Athenian notions of democracy [27].

5.2 How Should We Learn from Mistakes?

The Behavioural Economics Team of the Australian Government (BETA) represent another relatively new phenomenon in some governments—they exist to apply behavioural insights into contemporary policy problems, often using the tactic of making small changes or 'nudges' to address longstanding problems. While more seasoned policy scientists might see this as a modern version of Lindblom's incrementalism, in this manifestation one of the more interesting elements of the work of the BETA within the Department of Prime Minister and Cabinet is their stated commitment to learning from mistakes and to publishing the results of their work, even if it shows that particular interventions were not as successful as hoped for. While there are no readily available assessments of how this principle is applied in practice, it represents a commendable position for any government entity to take.

In a similar vein, some governments—national as well as state or provincial—have created new positions, such as Government Chief Scientist, Chief Economist or Government Statistician to oversee and ensure the integrity of research carried out by or on behalf of the government. In the UK, the Government Office for Science published recently Guidance for Government Chief Scientific Advisers and their Officials to clarify roles and responsibilities.

5.3 Can and Should Evaluation Researchers Speak Truth to Power?

In many OECD countries, especially those that adhere to the Westminster system of government, elected politicians are supported in their roles by professional public servants and while under threat in recent years, this public or civil service has been relatively permanent. The character of Sir Humphrey Appleby in the BBC TV satirical comedy Yes, Minister, is a Permanent Secretary, meaning he remains while the ministers he serves come and go according to electoral cycles and factional politics within the parties of government. The underlying logic of this arrangement is that these public or civil servants are able to offer 'frank and fearless' advice to their political masters without having to worry that they might lose their jobs by suggesting to a minister that his or her preferred policy has some logical flaws or serious impracticalities and is unlikely to achieve its objectives.

This approach of being frank and fearless or speaking truth to power remains an important part of the discourse of government, even as evidence mounts that the public service in Australia and elsewhere are being systematically politicised to the detriment of sound policymaking and good outcomes [28].

But of course, the assumptions of professional objectivity and the commitment to speak truth to power have also been subject to sustained criticism and indeed the very success and popularity of Yes, Minister lay in its portrayal of Sir Humphrey and his fellow mandarins as powerful people in their own right, with policy agendas of their own, even if these are small-c conservative and not especially partisan. Equally insightful and funny portrayals of the relations between politicians and their public servants in Australia can be seen in the ABC TV programs The Hollowmen and Utopia, programs that some public servants have said are far too close to their day-to-day worlds to be watchable.

The point of referring to these popular portrayals of the relationships between public servants who are democratically elected and those who are appointed because of their professional skills and scientific or technical knowledge is to provide some context for the environment in which evaluators, urban analysts and urban researchers go about their work. Some might in fact be public servants working in one of the levels of government, but many will be one step removed and work on commission for those public servants. Whatever the precise relationship, and whether the researcher/analyst works in or for government on commission, the political context of any piece of analysis or research should be recognised and acknowledged.

6 Conclusions

To the extent that urban analysis is a form of evaluation and is something done by practitioners of planning, broadly defined, then evaluation research is one of the core practices and competencies of urban planners. Whether they are preparing plans and policies or assessing development proposals in the light of these plans, planners are engaged, unavoidably, in the practice of evaluation. However, just as M. Jourdain in Moliere's Le Bourgeois Gentilhomme did not realise he had been speaking prose all his life, many planners might not realise they have also been practising evaluators. Of course, they might not choose to add this to their CV, but typically it pays to be aware of some of the key principles of evaluation outlined above as this will also, typically, make one a better planner.

Even if professional planners or evaluation researchers are not themselves formally responsible for making decisions about development proposals, strategic planning options or the continuation of programs of intervention, we still tend to hold to the idea(l) that formal decision-makers will pay attention to the professional assessments and evidence presented by planners and evaluators. This is not to say that formal decision-makers—duly elected politicians for example—always accept the advice of their professional planners, evaluators or urban analysts, not least because decisions require the application of political as well as scientific or professional judgement. But if decisions are made that fly in the face of substantial bodies of evidence and professional or scientific advice, then politicians might be held to account and even rejected by their constituents at future elections.

Evaluation research is all about values, goals and objectives—they serve as the benchmarks or yardsticks against which we collect and apply evidence. This is not to say that researchers must accept or agree with these values, simply that we should use these as yardsticks in assessing the success or failure of an intervention or a policy program. We have seen though that success or failure is rarely a simple binary choice. While not impossible, it is unlikely that a policy or intervention is an unequivocal failure in all its dimensions, or indeed a success across the board. Like the curate's egg, it is more likely to be good in parts. Perhaps the greatest challenge facing evaluation researchers lies in avoiding the conclusion of uncertainty; that we cannot say one way or another whether an intervention is successful or not and that further research is necessary. This is sometimes what distinguishes evaluation researchers working as private consultants from those who remain wedded to academic principles: the consultant evaluator will deliver a clear conclusion on time, whereas academic evaluators have been known to take a long time failing to reach a conclusion.

As in many other fields of research, there are craft skills to be developed and applied in evaluation research and practitioners should be looking to continuously improve these. Experience in practising this craft should also bring a heightened awareness of the politics of evaluation research. This might manifest itself in being able to convey conclusions in ways that capture the complexity of the enterprise without lapsing into ambiguity or inconclusiveness, or that delivers a message of relative failure in ways that do not antagonise the champions of a policy. This is not to say that uncomfortable truths should be avoided, merely that there are ways of speaking truth to power that increase the chance of powerful listening.

Evaluation research is an essential component of urban analysis, broadly defined, and can provide the foundation for better planning and the delivery of better planning outcomes. Like all applied research its practitioners face a number of significant challenges in designing and delivering high-quality evaluation research, including the conceptualisation of what constitutes high-quality work. Because one of the important foundational principles of evaluation research is to speak truth to power, even if the powerful do not always appreciate being told their favourite policy measure is not succeeding in its own terms, it is not always appropriate to judge the quality or value of a piece of evaluation research on the basis of its acceptance—enthusiastic or otherwise—by those commissioning it. But, by understanding and applying its foundational principles and by developing and applying the craft skills of good research, the evaluation practitioner will at least know they have done their best in the challenging environment of evaluation research.

Key Points

- Urban analysis covers many fields, from the assessment of development proposals through research on poverty and inequality in towns and cities to studies of the impact of urban policy measures.
- Evaluation research is an important element of urban policy analysis and involves a range of research techniques and approaches. Few, if any of these research techniques and approaches relate only to urban phenomena or to evaluation work.
- The philosophy, theory and practice of evaluation are subject to constant debate and competing paradigms of evaluation wax and wane. At present, realistic evaluation offers a compelling account of how to design and conduct evaluations that are empirically rigorous and conceptually plausible.
- Evaluation research necessarily involves making judgements about the value or success of something. These judgements are based on criteria, which are typically not determined by evaluators, but by the designers of policies and interventions.
- Making judgemental assessments, even if others have formal responsibility for accepting (or not) and applying them, is part of an inherently political process. Good evaluators recognise this and try to manage it without making unfounded claims to be 'above politics'.
- The practice of evaluation presents a number of ethical as well as political challenges with the obligation to present truthfully the results of a rigorous evaluation being one of the most important, regardless of the aspirations of those commissioning the evaluation. 'Speaking truth to power' remains a fundamental element of good evaluation practice, but remains also a complex principle to apply in practice.

Further information

For those wanting to get further information about evaluation research, there are a number of useful suggestions for further reading:

- Alan Clarke and Ruth Dawson (1999) Evaluation Research: An Introduction to Principles, Methods and Practices, London: Sage
- David Taylor and Susan Balloch (2005) The Politics of Evaluation: Participation and policy implementation, Bristol: The Policy Press
- Pawson, R. (2006) Evidence-based policy: A Realist Perspective, London: Sage
- Colin Robson (1993) Real World Research: a Resource for Social Scientists and Practitioner-Researchers

References

1. Patton MQ (2020) Evaluation use theory, practice, and future research: reflections on the alkin and king aje series. Amer J Eval 41(4):581–602
2. Kemp S (2012) Evaluating interests in social science: beyond objectivist evaluation and the non-judgemental stance. Sociology 46(4):664–679
3. Scriven M (1991) Prose and cons about goal-free evaluation. Eval Pract 12(1):55–62
4. HY R (1978) Some aspects of symbolic education policy: a research note. In: The Educational Forum. Taylor & Francis
5. Boaz A, Davies H (2019) What works now?: evidence-informed policy and practice. Policy Press
6. Newman J, Cherney A, Head BW (2016) Do policy makers use academic research? Reexamining the "two communities" theory of research utilization. Public Adm Rev 76(1):24–32
7. Shaxson L, Boaz A (2020) Understanding policymakers' perspectives on evidence use as a mechanism for improving research-policy relationships. Environ Educ Res, 1–7
8. Slater T (2006) The eviction of critical perspectives from gentrification research. Int J Urban Reg Res 30(4):737–757
9. Harré R, Srivastava N (1983) An introduction to the logic of the sciences. Springer
10. Bhaskar R (2013) A realist theory of science. Routledge
11. Pawson R, Tilley N (1977) Realistic evaluation. Sage
12. Campbell DT (1971) Reforms as experiments. Urban Affairs Quart 7(2):133–171
13. Campbell D, Stanley J (1963) Experimental and quasi-experimental evaluations in social research. Rand Mc-Nally, Chicago
14. Oakley A (2000) Experiments in knowing: gender and method in the social sciences2000
15. Guba EG, Lincoln YS (1989) Fourth generation evaluation. Sage
16. Rossi PH, Freeman HE, Wright SR (1985) Evaluation: a systematic approach. Beverly Hills. CA: Sage
17. Chen H-T, Rossi PH (1983) Evaluating with sense: the theory-driven approach. Evaluat Rev 7(3):283–302
18. Chen HT (1990) Theory-driven evaluations. Sage Publications
19. Robson B, Bradford M (1995) An evaluation of urban policy. Paul Chapman Publishing Urban Policy Evaluation, London
20. UN-HABITAT, Monitoring and Evaluating National Urban Policy: A Guide (2020) UNHABITAT: Nairobi

21. Burton P, Goodlad R, Croft J (2006) How would we know what works? Context and complexity in the evaluation of community involvement. Evaluation 12(3):294–312
22. Burton P (2009) Conceptual, theoretical and practical issues in measuring the benefits of public participation. Evaluation 15(3):263–284
23. Rossi PH, Lipsey MW, Freeman HE (2004) Evaluation: a systematic approach. Sage Publications
24. Burton P (2007) Modernising the policy process: making policy research more signficant. Policy Stud 27(3):173–195
25. UNESCO, Recommendation on science and Scientific Research. 2018, UNESCO: Paris
26. Wildavsky AB (1989) Speaking truth to power. Transaction Publishers
27. Robinson EW (2008) Ancient Greek democracy: readings and sources. Wiley
28. Head BW (2013) Evidence-based policymaking–speaking truth to power? Austr J Public Administr 72(4):397–403

Communicating Urban Research

Stephanie Wyeth and Laurel Johnson

Abstract Urban research influences urban policy, practice and public opinion. When it is effectively communicated, urban research energises, excites and opens new ways of looking at the urban environment to solve its many complex challenges. Good communication of urban research adds to the body of academic knowledge and it creates change in our cities, environments and communities. Whether you are an academic researcher, a consultant, a student or an advocate or activist, your research can influence policy, practice and opinion, but it must be effectively communicated to do so. This chapter sets out a 'how to' guide for the effective communication of urban research. This is practical advice for urban researchers and students to build their academic and professional profiles and contribute to urban knowledge and practice. The guidance in this chapter is based on the author's collective experience as influential urban researchers and urban strategists in both public (Government and University) and private (consulting) sectors. The chapter overviews the context for urban research and then follows a basic sequence of *why, when* and *how* to communicate urban research. In the latter part of the chapter, we offer eight communication tips and practice notes to maximise the impact of your research.

1 The Context for Urban Research

Our cities and communities are facing complex and challenging issues across economic, social and environmental themes. At the same time, an avalanche of technological, civic and political factors are shaping how urban information, data and research is conceived, commissioned, reported and interpreted. These factors include:

S. Wyeth (✉) · L. Johnson
University of Queensland, Brisbane, QLD, Australia
e-mail: s.wyeth@uq.edu.au

- the shifting role of 'the expert' in public discourse, and the rise of the 'influencer' or 'community voice' in planning and design processes [1];
- concern regarding the relative independence of academic research and the growing influence of private, political and commercial interests on research and public policy;
- a growing distrust of government, institutions, mainstream media and the rise of 'fake news' [2];
- the exponential growth in the amount of detailed and personalised information and open data generated through daily interactions, and the ability to use this information to shape and inform decisions [3]; and
- the rise of analytics and visualisation platforms such as geographic information systems (GIS), story mapping and user generated data often without adequate time to process, reflect or check for accuracy or bias [4].

This context means that there is a plethora of information and data on all aspects of the urban environment. A key role for the urban researcher is to filter, analyse and make sense of this information so that it is effectively packaged and communicated to disparate audiences. This is important to ensure the distribution of information from ethical and trustworthy sources.

2 Why Communicate Urban Research

There are three connected reasons to communicate urban research. First, to elicit one of the five audience responses identified in the figure below. Second and related to this, to enlist support to solve complex urban problems and third, to fairly share knowledge of urban issues.

2.1 Elicit Audience Response

To be effective, the communication of urban research will promote one or more of the following audience responses: care and motivate, understand and inform, remember, consider and act, and advocate and share (see Fig. 1). To evoke any of these responses, the urban researcher needs a clear understanding of the audiences for research and to communicate to them through multiple media.

2.2 Solve Complex Urban Issues

There is a sense of urgency to tailor and communicate urban research to address the damaging impacts of human-induced climate change and create healthy and

Fig. 1 Desired impact and
response to urban research.
Adapted from [5]

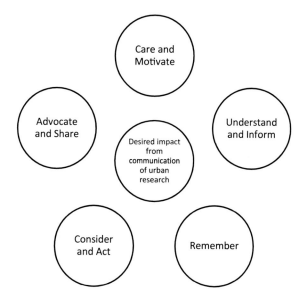

safe post-pandemic cities. These matters are amongst other global megatrends and
disruptions that need the attention of urban researchers. For those communicating
for a professional purpose, your analysis could kick-start a new initiative, inform a
policy or planning process or secure support for specific interventions and projects.
Urban analysis can also become a lightning rod for concerned citizens and advocates
seeking to affect change in a city or region. For those communicating for an academic
purpose, your research has the potential to create knowledge, enable discovery and
influence decision-makers and public opinion [6].

2.3 Fairly Share Knowledge of Urban Issues

Urban researchers have responsibility to conduct research that contributes to
sustaining ecological systems and liveable environments. They also have a responsibility to ensure that their research is accessible, promoted, shared equitably and
communicated to multiple audiences.

3 When to Communicate Urban Research

The urban environment is dynamic and priority issues and trends change quickly.
Urban researchers must communicate their analysis and expertise when it is needed.
For the University-based urban researcher, this dynamism brings tension as they

balance the University research timeline with the urgency for knowledge to influence current urban practices and opinion. Researchers can't wait until the research project is complete to communicate the research questions, enlist support or release key findings or commentary. The timely communication of urban research is essential to its effectiveness.

There is a plethora of communication mediums that provide the basis for the communication of urban research to influential audiences. These include:

- social media platforms;
- print media (local, city-wide, State and National);
- radio (radio shows, expert commentary);
- urban researcher networks;
- journals with academic and non-academic readership;
- urban planning practice networks;
- professional associations for urban planners, designers and allied professions; and
- urban development and city leadership and advocacy group web-sites.

4   How to Effectively Communicate Urban Research

There are multiple audiences for urban research and the most impactful urban research is communicated to these different audiences through many media.

So how do you do this—as a student, researcher, advocate, or planning practitioner? We have set out eight simple tips and provide practice notes to support you to effectively communicate your research:

1. Practice ethical research
2. Understand power and influence
3. Present the complete picture
4. Plan your communication
5. Illustrate and illuminate
6. Write clearly
7. Be informed of trending urban issues
8. Be relevant.

4.1   Practice Ethical Research

No matter what the motivation for the research is, it is important to understand human research ethics and act with integrity in designing, conducting and reporting research. Acting ethically supports researcher's credibility and future opportunities. If you are uncertain or unclear about a research project, method or finding, say so. Sometimes, research can be derailed or produce results that are not anticipated at the outset. In these cases, communicate these limitations to the audience.

Practice Note: Reflect on ethics and integrity

Your reputation as an ethical urban researcher is important for brokering new funding, clients and collaborators and to promote your role as a trusted expert. Whether research is for a private client or a public audience (or both), your professionalism and ethics are foundational to your research practice and reporting. It is assumed that all urban researchers undertake research according to principles and practices of ethical human research.

In Australia, the National Health and Medical Research Council (NHMRC) is a Government body that provides authoritative information on the principles and practices of ethical human research. These resources are available for Australia's academic and non-academic urban researchers.

4.2   Understand Power and Influence

Urban analysis is commissioned, designed, accessed and read by researchers, planners, designers, policymakers, urban activists and those interested in urban life and the shape of our cities and towns. This research is often triggered through a thoughtful observation of, or dialogue on a current or emerging pattern, experience or event. Any results are used as a 'point of truth' to better understand an urban issue or challenge or validate an investment or policy position.

Urban research should be considered as part of an ongoing dialogue on place or community which has value and aspirations, assets, diversity of cultures, views, opinions and experiences. Write for a diverse audience and consider a writing and graphic communication style that is personal, authentic and informative. Avoid the use of acronyms and other technical terms in both academic and non-academic media. A good communicator will take the time to explain the key concepts and ideas to ensure that the reader is able to interpret and form an opinion based on the research.

Increasingly, the results of urban analysis are used to simultaneously engage academic and technical audiences as well as the public. Unfortunately, urban analysis findings (like all research) can be exclusionary and risk being politicised, miscommunicated and misunderstood if the writer has not taken the time to consider who may be impacted by the research or the planning interventions that follow.

Practice Note: Avoid technical language and assumed knowledge

There are many instances in academic papers and professional reports where the reader is expected to have a level of knowledge and this excludes many audiences from accessing your research findings. This reinforces and exacerbates power structures that exclude community members and other non-academics from participating in deliberations about their suburbs and cities.

4.3 Communicate the Complete Picture

Understand the context and define and explain terms in your research field. Presenting your research without describing key features of the economic, social, political and environmental setting can adversely impact your audience's engagement with your findings. This is critical in urban research as there is a risk of practitioners and policymakers jumping to a 'one size fits all' solution based on your research with out considering local conditions. Examples of this approach could include:

- Creating a profile of city cycle infrastructure without taking into account cultural and equity barriers to bike ownership and access, impact of climatic conditions and local topography, and school travel policies;
- Recommending a city-wide solution to on-street car parking without completing a comparative assessment of neighbourhood residential density, public transit system quality or location of employment nodes.
- Evaluating the social and economic impacts of a natural disaster response strategy for a single event, without taking into account the cyclical nature of natural disasters in a location.

Practice Note: Understand and explain the context and define common terms

Urban researchers must communicate the context, common terms and the complete picture. This context setting can be visual and engaging and explaining terms can also be done visually.

Practice example: Mapping the pandemic

The coronavirus pandemic 2020–2021 provides an excellent backdrop for understanding the benefits and risks of communicating the context for urban themes and issues. In this example (see Fig. 2), a series of short slides are used to illustrate the emerging body of knowledge on the relationship between cities, planning, housing and COVID-19. The research challenge is introduced. Density is defined and explained so that the audience understands the difference between density and overcrowding as factors in the spread of the pandemic [7].

A direct quote is used to describe the lived experience of the pandemic, and an infographic is inserted to highlight the spatial distribution of the impacted population, relative to the local government area [8]. The presentation slides close with an example of a program response, to highlight the potential to take targeted action in a different planning context on this critical public health issue.

4.4 Plan Your Communication

When planning to communicate urban research, ask yourself these questions:

- Who is the audience for this research?

- Which print and social media outlets and forums will I target to reach these audiences?
- Will the research be freely available online to any audience?
- Can the research be read on a mobile device and if printed in black and white?

Practice Note: Consider format, structure, style and visualisation

Design written and visual communication to engage the reader. Curate the elements of your communication to maximise legibility and interest and navigate the reader. The example below shows the careful use of graphics and text to build the reader's knowledge and interest (Fig. 3).

Slide 1 - Emerging Research

COVID-19 is a systemic disease, and cities are large complex systems. Data rich analysis and research is required to understand what is happening before developing a systemic response to the issue.

Similar to responses to climate change, digital transformation etc

Slide 2 - Emerging Research

Does Density Aggravate the COVID-19 Pandemic?

"The impact of density on emerging highly contagious infectious diseases has rarely been studied.

In theory, dense areas lead to more face-to-face interaction among residents, which makes them potential hotspots for the rapid spread of pandemics.

On the other hand, dense areas (urban areas) may have better access to health care facilities and greater implementation of social distancing policies and practices."

✓ Density doesn't always = higher infection rates as residents of dense places are more likely to practice basic social distancing than their counterparts in suburban and exurban areas.

✓ Larger metropolitan areas have significantly higher rates of infection – spread through the movement and interaction of people (connectivity vs density).

✓ Socio-economic factors within the population – age, health, employment, housing, ability to access 'remote working' and social distancing options

Shima Hamidi, Sadegh Sabouri & Reid Ewing (2020): Does Density Aggravate the COVID-19 Pandemic?, Journal of the American Planning Association, DOI: 10.1080/01944363.2020.1777891

Fig. 2 Communicating urban themes in the early days of the 2021 pandemic from Wyeth, S. (2020). Introductory session on health and planning issues, PLAN1100 Foundational Ideas in Planning (course for first year planning students), The University of Queensland

Slide 3 - New York

"In Manhattan you might have only two people in a studio apartment, and in parts of Brooklyn or Queens you might have a family of five or six people in a room that size.

Mortality rates were higher in neighborhoods with lower incomes and less density across the geographic space but more density in a given home. "

Quote from article by Rogers (2020), Wired

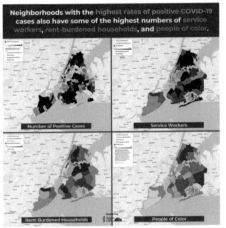

Slide 4 - Emerging Research

Why does COVID-19 incidence and spread vary across and within cities.

Some emerging factors:

✓How quickly public health measures were implemented? Limiting people's movement through school closures, home isolation/working from home, social distancing, curfews.

✓Number of people per unit of housing (key social determinant) – close contact is critical.

Slide 5 - Emerging Research

There is a strong causal link between overcrowding, poor quality housing and settlements and the incidence and prevalence of disease.

Fig. 2 (continued)

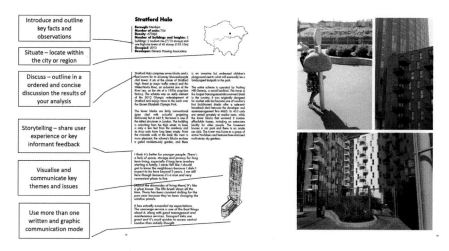

Fig. 3 Effective communication of research findings (mixed method approach) [9]

4.5 Use Multiple Mediums to Illustrate and Illuminate

Long written reports and articles are possibly the least engaging but most common form of communication in urban research. This is unfortunate. To illuminate research, communication should be short, interesting and include relevant and clear visualisation.

The urban researcher's tool kit includes visualisation tools that achieve a number of purposes. These include:

- enhance and inform multiple urban planning roles and tasks such as plan making, regulation, mediation, advocacy, technical analysis;
- facilitate participatory and communicative planning processes;
- integrate and interpret data from disparate sources and in a variety of formats; and
- facilitate iterative processes in the analysis and recursive exploration of data [1].

Practice Note: Visualisation interprets and communicates an idea to make it understood

Well-crafted visuals can evoke any of the five audience responses of: care and motivate; understand and inform; remember; consider and act; and advocate and share. Table 1 presents common visualisation techniques currently used by Australian urban researchers, urban policy and planning authorities and their benefits and challenges.

In our experience, icons are widely used to communicate complex ideas. Practice example 2 presents an example of communicating the role of the planner. This is an example of visualisation through icons and diagrams in the presentation of complex information. This style is common in technical presentations and also highly suited to the communication of urban research. Consider the use of animations, movement, music, video and colour in visual presentations of your research.

Table 1 Common visualisation tools in urban research (adapted from [1])

Visualisation	Examples	Benefits	Challenges
1. Graphics	Sketches, diagrams, photographs, maps	Supports the clear communication of research methods, context, findings and recommendations	Must be accompanied by text to interpret the messages that are being communicated in the graphics. Must be meaningful and add value to the communication of the research
2. Geographic information systems (GIS)	Mapping, analytics, data management, Story Maps	Can be static, dynamic or interactive and interacts with many apps Used to gather and share information and knowledge Useful in participatory planning process to test and communicate scenarios. Can combine text, maps and other multimedia in one presentation (storymaps) Crowdsource community observations	Need some level of technical skill to use GIS. Can overwhelm the audience so they need to be simply presented and may need interpretation. Crowdsourced data may not be accurate
3. Dimensional (3D) models and simulations	City Engine, Fly through, virtual reality apps	Can generate and share multiple design scenarios that look realistic and show the impacts of development and change. Can communicate a vision for a place. Can visualise, experience and measure the impact of scenarios	Need some level of technical skill and access to data to create the 3D models though can generally be viewed on devices such as smartphones
4. Video, websites, power point (PPT) presentations	Dashboards, short movies, PPT presentations	Dashboards present multiple measures in one place and can show results over time for many variables at multiple locations. Short movies can be shared to social media. PPT presentations when well crafted, will discipline the researcher to present the key findings and insights and can be shared to multiple audiences	Need reliable and valid dashboard data and resources to build and maintain dashboards Short movies need to be thoughtful and carefully edited to engage audiences. PPT presentations must follow rules such as maximum three points per slide, more graphics, less words and others

Practice example 2: Use icons and engaging images to clarify complex ideas

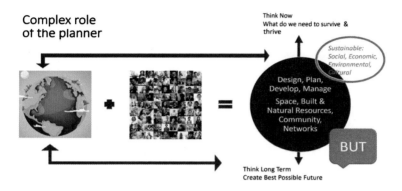

4.6 Written Communication Should be Clear and Accessible

Academic urban researchers are not generally taught how to write an effective paper, design and give a verbal presentation or create a video. In order to be published, academics usually follow the conventions of peer-reviewed publications. Many academic publications are difficult to navigate and hence poorly communicated. Academic writing often overlooks the simple conventions of written English, such as the subject-verb sequence. This results in the unclear communication of important urban research and this limits its audience and impact [6].

Our advice is to make all communications engaging and easy to understand. This advice applies to academic and non-academic reporting.

Practice Note: Write clearly and concisely

We encourage researchers to adopt the following approach to their writing (details in Table 2) whether the audience is academic or non-academic. Much academic writing is not 'good writing' as it does not clearly communicate information. Table 2 is adapted from an authoritative guide to writing English.

4.7 Be Informed of Trending Urban Issues

To maximise their influence, urban researchers are aware of contemporary urban issues and are ready to share their analysis and commentary on these issues with multiple audiences.

Practice Note: Use academic and non-academic forums to share your research and insights.

Table 2 Effective english writing tips [10]

Effective Writing	Description	Examples
1. Follow a Character–Action Sequence	Good writing has identifiable characters as subjects and verbs that express action. NOTE: Characters can be inanimate such as data, techniques, plans, places and others. Characters (or subjects) are not always people. Unclear writing has missing characters and empty verbs	**Students** (character) in planning **write** (action) clearly because **they** (character) **attend** (action) writing workshops
	Good writing has clear characters in the subject position at the start of the sentence Good writing uses verbs that express strong actions Good writing uses active verbs	
2. Avoid nominalisation of Verbs	Sometimes, verbs become 'empty verbs' or nouns: – discuss and discussion, evaluate and evaluation, analyse and analysis – evaluation, analyse, analysis AVOID NOMINALISATION = When a doing word becomes a naming word	Example 1. We failed to analyse the instructions, so we did not improve NOMINALISATION—Our failure to conduct analysis of the instructions led to a lack of improvement Example 2. We clearly improved. NOMINALISATION—There was a clear improvement
3. Use subordinating conjunctions	Subordinating conjunctions can help you move ideas around and avoid nominalisation Examples are although, because, how, if, provided that, since, so long as, so that, that, though, unless, when Try to use subordinating conjunctions instead of nominalising your verbs	
4. Good sentences	Have clear characters doing strong actions Clear characters are in the subject position at the start of sentences Use verbs that express strong actions. Good sentences are short sentences. One sentence should not have more than 20 words	

(continued)

Table 2 (continued)

Effective Writing	Description	Examples
5. Ordering Ideas	Have an order to your writing such as chronological or numerical or logical Start with an idea that is familiar to your reader—a shared and agreed experience or fact	For example, "Of the three issues identified on the field trip, the first is …" Navigate the reader through your writing. Clear writing helps the audience to understand the point/s that you are making

Influential Australian urban researchers contribute to the high-profile public-facing outlets. At the time of writing, those outlets included:

- The Conversation;
- The Guardian;
- The Urban Developer;
- LinkedIn groups with an urban interest; and
- Urban planning, urban policy and other urban-related social media forums.

Forums and distribution outlets change, and more are appearing so invest time to scan and use the multiple platforms that deliberate contemporary urban issues. Make yourself and your expertise known to the University media and marketing sections of your institutions and introduce yourself to journalists who cover urban issues.

Urban researchers should participate in urban advocacy forums such as 'Committees for…' events (each Australian city has a 'Committee for…') and join the State urban planning professional association/s such as the Planning Institute of Australia and urban design associations and good design advocacy bodies.

4.8   Be Relevant

Urban researchers need to be agile and work with several roles, tools and methods to ensure the communication of research is timely and trusted.

Practice Note: Stay up to date with urban trends and promote yourself and your research

Regularly communicate your research interests and insights. Keep your LinkedIn profile up to date with your latest publications, media commentary and current research interests. Make regular contributions to relevant LinkedIn 'urban' forums and groups and post and comment regularly. Include your research interests and links to your latest/high impact publications and LinkedIn and research profile links in your email sign-off.

5 Conclusion

When it reaches many audiences, urban research influences urban planning practice and urban policy. The effective communication of urban research maximises its reach to those audiences. While peer-reviewed, subscription-based academic journals remain relevant to the distribution of academic research, they are not central outlets for the influential urban readership. To maximise impact, the urban researcher needs to use the many forums available to distribute their research, understand their audiences, be timely, ethical and relevant and effectively write and visualise their research.

Key Messages:

- When effectively communicated, urban research influences urban policy, practice and opinion and engages and motivates audiences
- Urban researchers have a responsibility to effectively communicate their research to academic and non-academic audiences
- Urban researchers analyse and communicate information to multiple audiences
- For maximum impact, urban research is communicated at the right time, in multiple media
- Urban researchers must stay informed of current urban trends and issues
- Visualisation is essential to the effective communication of urban research.

Further information

For those wanting to get further information about effective communications see:

Ann Esnard (2012). Visualising Information. In Crane, R., & Weber, R. (eds). *The Oxford Handbook of Urban Planning*. Oxford Handbooks Online.

or

Jenny Rankin (2020). *Increasing the Impact of Your Research: A Practical Guide to Sharing Your Findings and Widening Your Reach*. Routledge: New York.

or

Cody Weinberger, James Evans & Stefano Allesina, (2015). Ten Simple (Empirical) Rules for Writing Science. *PLOS Computational Biology*. 11(4).

or

Joseph Williams, & Joseph Bizup, (2014). *Style: Lessons in Clarity and Grace*. Pearson: Boston.

References

1. Esnard A-M (2012) Visualizing information, in The oxford handbook of urban planning
2. Lazer DM et al (2018) The science of fake news. Science 359(6380):1094–1096
3. Pingo Z, Narayan B (2016) When personal data becomes open data: an exploration of lifelogging, user privacy, and implications for privacy literacy. In: International conference on asian digital libraries.Springer
4. Lee CS et al (2016) Investigating the use of a mobile crowdsourcing application for public engagement in a smart city. In: International conference on asian digital libraries. Springer
5. Rankin JG (2020) 4 Pages of research dissemination secrets. increasing the impact of your research: a practical guide to sharing your findings and widening your reach, p. N/A–N/A
6. Freeling B, Doubleday ZA, Connell SD (2019) Opinion: how can we boost the impact of publications? Try better writing. Proc Natl Acad Sci 116(2):341–343
7. Hamidi S, Sabouri S, Ewing R (2020) Does density aggravate the COVID-19 pandemic? Early findings and lessons for planners. J Amer Plann Assoc 86(4):495–509
8. Dickson E (2020) New map shows COVID-19 is hitting people of color hardest. Rolling Stone. Retrieved from https://www.rollingstone.com/culture/culture-news/covid19-coronavirus-pandemic-low-income-people-new-york-city-976670
9. Scanlon K, Blanc F, White T (2020) Living in a denser London: how residents see their homes
10. Williams JM, Bizup J (2014) Lessons in clarity and grace. Pearson

Printed in the United States
by Baker & Taylor Publisher Services